Institutionalizing Monitoring and Evaluation in Africa's Public Sector: Challenges and Best Practices

Elimon Nyamambi

Published by Elimon Nyamambi, 2025.

INSTITUTIONALIZING MONITORING AND EVALUATION IN AFRICA'S PUBLIC SECTOR: CHALLENGES AND BEST PRACTICES

First edition. August 4, 2025.

ISBN: 979-8231828166

Written by Elimon Nyamambi.

This book is dedicated to the countless M&E practitioners, civil servants, and community leaders across Africa who work tirelessly to ensure that government policies and programs deliver on their promises. It is for those who, despite facing systemic challenges and limited resources, remain committed to the pursuit of credible data, honest evaluation, and a culture of continuous learning. Your resilience, integrity, and dedication are the true foundation of a thriving M&E ecosystem, and your work is the living embodiment of the principles discussed within these pages.

Above all, this work is dedicated to the people of Africa. It is our shared hope that through the power of rigorous M&E, we can build a continent that is not only prosperous and peaceful but also transparent, accountable, and just. May this book serve as a tool in the collective effort to turn the ambitious goals of Africa's Agenda 2063 and the Sustainable Development Goals into a tangible reality for every citizen.

"Until the lion learns how to write, every story will glorify the hunter."— *African Proverb*

Institutionalizing Monitoring and Evaluation in Africa's Public Sector: Challenges and Best Practices

By
Elimon Nyamambi
(M&E SPECIALIST)

DEDICATION

This book is dedicated to the countless M&E practitioners, civil servants, our governments, and community leaders across Africa who work tirelessly to ensure that government policies and programs deliver on their promises. It is for those who, despite facing systemic challenges and limited resources, remain committed to the pursuit of credible data, honest evaluation, and a culture of continuous learning. Your resilience, integrity, and dedication are the true foundation of a thriving M&E ecosystem, and your work is the living embodiment of the principles discussed in this book.

Above all, this work is dedicated to the people of Africa. It is our shared hope that through the power of rigorous M&E, we can build a continent that is not only prosperous and peaceful but also transparent, accountable, and just. May this book serve as a tool in the collective effort to turn the ambitious goals of Africa's Agenda 2063 and the Sustainable Development Goals into a tangible reality for every citizen.

ABBREVIATIONS

AfrEA: African Evaluation Association

AUC: African Union Commission

AUDA-NEPAD: African Union Development Agency - New Partnership for Africa's Development

CLA: Collaborating, Learning, and Adapting (USAID)

CLEAR: Centres for Learning on Evaluation and Results

CSOs: Civil Society Organizations

DPME: Department of Performance, Monitoring and Evaluation (South Africa)

GWM&E: Government-Wide M&E

IDS: Institute of Development Studies

ILO: International Labour Organization

JLMP: Joint Labour Migration Programme

M&E: Monitoring and Evaluation

NIMES: National Integrated Monitoring and Evaluation Strategy (Uganda)

NGOs: Non-Governmental Organizations

PPMED: Policy Planning, Monitoring and Evaluation Division (Ghana)

PMP: Performance Management Plan

PRSC: Poverty Reduction Support Credit (Uganda)

RBM: Results-Based Management

RCTs: Randomized Controlled Trials

RESILIM-O: Resilience in the Limpopo (Olifants Basin) Program

SANBI: South African National Biodiversity Institute

SANParks: South African National Parks

SDGs: Sustainable Development Goals

Sida: Swedish International Development Cooperation Agency

UNDP: United Nations Development Programme

USAID: United States Agency for International Development

Introduction

The Global Context: M&E and the Sustainable Development Goals (SDGs)

The 21st century has been defined by an unprecedented commitment to global development goals. The Sustainable Development Goals (SDGs), adopted by all United Nations Member States in 2015, represent a universal call to action to end poverty, protect the planet, and ensure that all people enjoy peace and prosperity by 2030. Unlike their predecessor, the Millennium Development Goals (MDGs), the SDGs are more expansive, complex, and interconnected. They address a wide range of social and economic development issues, from climate action and clean energy to gender equality and quality education.

Achieving these ambitious goals requires more than just good intentions; it demands a robust system for tracking progress, identifying what works, and course-correcting when things go astray. This is where Monitoring and Evaluation (M&E) becomes not just useful, but absolutely essential. M&E provides the tools to measure progress towards specific targets and indicators. It helps governments, organizations, and communities understand whether their policies and programs are delivering the intended results and whether they are doing so efficiently and equitably.

Globally, M&E has evolved from a donor-driven requirement into a fundamental pillar of good governance and effective public management. It is the mechanism through which nations can hold themselves accountable to their citizens and to the international community for their commitments under the SDGs. Without reliable M&E systems, it would be impossible to answer a fundamental question: are we on track to achieve the SDGs by 2030?

The African Context: M&E as a Tool for Development and Governance

While the global context of the SDGs provides a backdrop, the need for institutionalizing M&E in Africa's public sector is particularly acute and rooted in a unique set of challenges and opportunities. For decades, many African nations have grappled with issues of poverty, inequality, and weak public institutions. Public funds, often scarce, have been mismanaged, and well-intentioned development programs have failed to deliver lasting impact.

In this environment, M&E is a powerful tool for transformation. It moves governance away from a focus on simply spending money or implementing activities and towards a focus on achieving tangible results. When properly institutionalized, M&E can:

- **Enhance accountability:** It allows governments to demonstrate to their citizens how public resources are being used and what is being achieved.
- **Improve decision-making:** By providing evidence on the performance of policies and programs, M&E helps leaders make more informed choices about where to allocate resources and what strategies to adopt.
- **Promote learning and adaptation:** M&E findings can be used to understand why certain interventions succeed or fail, enabling a culture of continuous improvement rather than repeating past mistakes.
- **Strengthen citizen engagement:** When M&E data is made public and accessible, it empowers citizens and civil society organizations to participate in governance and hold their leaders accountable.

The shift from a project-based approach to a systemic, government-led approach to M&E is not just a technical exercise; it is a political and cultural one. It signifies a move towards greater ownership of the development agenda and a commitment to transparency and results-based management.

Defining M&E: Beyond Mere Reporting

A common misconception is that M&E is simply about gathering data and writing reports. While these are components, a truly institutionalized M&E system goes much deeper.

Monitoring is the continuous, systematic process of collecting data on specific indicators to provide management and stakeholders with an indication of the extent of progress and achievement of objectives. It answers questions like: *Are we doing what we said we would do? Are we on track?*

Evaluation, on the other hand, is a more rigorous and periodic assessment of an ongoing or completed project, program, or policy. It seeks to determine the relevance, effectiveness, efficiency, impact, and sustainability of an intervention. Evaluation answers the more critical question: *Is what we are doing working? Why or why not?*

When we talk about institutionalizing M&E, we are not just talking about having a monitoring unit or conducting an occasional evaluation. We are talking about embedding M&E into the very DNA of public administration. This means:

- M&E is supported by clear legal and policy frameworks.
- It is adequately funded in national and sectoral budgets.
- It is staffed by competent and skilled professionals.
- Its findings are systematically used to inform policy, planning, and resource allocation.
- It is recognized by leaders and civil servants as a core function of government, not an optional extra.

The Scope and Purpose of the Book

This book is intended as a comprehensive guide for policymakers, public sector officials, development practitioners, academics, and students who are interested in or actively involved in the institutionalization of Monitoring and Evaluation (M&E) in Africa. The scope of our discussion will not be limited to the technical aspects of M&E—such as designing logical frameworks or collecting data—but will extend to the broader political, institutional, and cultural dimensions that are critical for long-term success.

The primary purpose of this book is to:

1. **Analyse the current state of M&E in Africa's public sector:** We will examine the progress made, the persistent challenges faced, and the evolving trends shaping the M&E landscape across the continent.
2. **Identify and articulate the core challenges:** We will provide a deep dive into the political, financial, technical, and cultural barriers that hinder the effective institutionalization of M&E. This includes a frank discussion about issues like a lack of political will, insufficient budgets, and a culture of resistance to accountability.
3. **Propose and discuss best practices:** Based on evidence and real-world examples, we will present a range of best practices and practical strategies for building robust and sustainable M&E systems. These will be grounded in the African context, acknowledging the unique socio-economic and political realities of the continent.
4. **Offer a framework for action:** The book will serve as a practical resource, providing a roadmap for governments and their partners to move from ad-hoc M&E activities to a fully integrated and results-oriented public administration system.
5. **Inspire a new generation of leaders:** Ultimately, we hope to inspire a new generation of public sector leaders to champion evidence-based decision-making and embrace M&E as a core function of good governance, thereby contributing to more effective development outcomes for their citizens.

We will draw upon a wide range of case studies from various African countries to illustrate key points and provide concrete examples of both successes and failures. This comparative approach will help readers understand that while challenges are often shared, solutions must be tailored to specific national and regional contexts.

Key Concepts and Terminology

To ensure clarity and consistency throughout the book, it is important to define some key terms we will be using.

- **Institutionalization:** This refers to the process of embedding M&E into the core functions, policies, procedures, and culture of a public sector organization or government. It moves M&E from a temporary, project-specific activity to a permanent and routine part of how the government operates.
- **Results-Based Management (RBM):** A management strategy that focuses on achieving specific, measurable, and tangible results. M&E is a fundamental tool for implementing RBM, as it provides the data and analysis needed to track progress towards those results.
- **Logical Framework (LogFrame):** A project management tool used to design, plan, and evaluate projects. It helps to clarify objectives, activities, indicators, and assumptions, providing a clear roadmap for what will be monitored and evaluated.
- **Theory of Change (ToC):** A comprehensive description and illustration of how and why a desired change is expected to happen in a particular context. It maps out the causal pathways from inputs and activities to short-term, medium-term, and long-term outcomes.
- **National M&E System:** A comprehensive, government-led system for monitoring the performance of national

development plans, policies, and programs. It typically involves various government institutions working together under a unified framework.

- **Public Sector:** Refers to all organizations and enterprises owned and operated by the government on behalf of the public. This includes ministries, departments, state-owned enterprises, and local government authorities.

These definitions will provide a common language as we delve into the more detailed chapters.

Chapter 1

The Evolving Landscape of Public Sector Management in Africa

Introduction

The narrative of Africa's public sector is a story of profound and ongoing transformation. For decades following independence, public administration on the continent was largely shaped by the legacies of colonial bureaucracy, characterized by rigid hierarchies, a top-down approach, and an overriding emphasis on process over results. While this model provided a framework for statehood, it often struggled to adapt to the complex challenges of poverty reduction, economic development, and nation-building. It was a model designed for control, not for fostering a dynamic and responsive public service.

This traditional paradigm has, however, been steadily giving way to a new and evolving landscape. The shift has been catalysed by a convergence of internal and external pressures. From within, a growing and more demanding citizenry, a vibrant civil society, and a new generation of political leaders are increasingly calling for greater accountability, transparency, and effective service delivery. From without, global development agendas, particularly the Sustainable Development Goals (SDGs), and the influence of international financial institutions have pushed for public sector reforms that prioritize efficiency, performance, and good governance.

1. Public Sector Reform in Africa and the Role of M&E

The drive to institutionalize M&E in Africa's public sector cannot be seen in isolation; it is a central pillar of the broader movement towards public sector reform that has swept across the continent since the 1980s. Faced with legacies of colonial administration, post-independence economic challenges, and the need to address persistent development issues, many African governments have embarked on comprehensive reforms to enhance the efficiency, effectiveness, and accountability of their public institutions.

Early waves of public sector reform, often influenced by structural adjustment programs, focused primarily on fiscal austerity and reducing the size of the civil service. While these efforts sometimes led to short-term economic stabilization, they often failed to improve the quality of public service delivery or address the root causes of inefficiency and corruption.

A new generation of reforms, however, has shifted the focus from mere cost-cutting to a more holistic approach centred on results-based management (RBM) and good governance. This is where M&E becomes indispensable. M&E provides the necessary tools and a crucial feedback loop for public sector reforms by:

- **Measuring the impact of reforms:** Is a new civil service code leading to better performance? Is a new anti-corruption agency reducing graft? M&E helps answer these questions with evidence.
- **Creating a culture of accountability:** By tracking and reporting on performance against targets, M&E systems hold public officials accountable for their actions and for the use of public resources.
- **Informing policy and program adjustments:** M&E data can reveal which reform initiatives are succeeding and which are not, allowing governments to adapt their strategies in real-time and avoid a one-size-fits-all approach.
- **Building trust with citizens:** A transparent and functional M&E system demonstrates a government's commitment to delivering on its promises, which is vital for building public trust and legitimacy.

In essence, M&E transforms public sector reform from a series of top-down directives into an evidence-based, results-oriented process. It shifts the emphasis from *what a government is doing* (e.g., passing a new law) to *what a government is achieving* (e.g., reducing poverty, improving health outcomes). Institutionalizing M&E is, therefore, not just an add-on; it is a fundamental mechanism for making public sector reform work in Africa.

1. **The shift from traditional public administration to results-based management (RBM).**

For much of the 20th century, public administration was dominated by a traditional, bureaucratic model. This model, rooted in the principles of a hierarchical structure, formal rules and procedures, and a clear division of labour, focused primarily on inputs and processes. Success was often measured by how well rules were followed, how much money was spent, and how many activities were completed. The central questions were: *Did we stay within budget? Did we follow the correct procedures?* The link between these activities and the actual impact on citizens' lives was often weak or entirely absent.

The limitations of this approach became increasingly apparent in the face of complex social and economic challenges. It led to a lack of innovation, inefficiency, and a public perception that the government was more concerned with its own internal workings than with delivering tangible value to the people it served.

This dissatisfaction, combined with global trends in new public management, catalysed a fundamental shift towards Results-Based Management (RBM). RBM represents a paradigm change, moving the focus of public administration from inputs and activities to outputs, outcomes, and impact. It asks a different set of questions: *What did we achieve for our citizens? Did our program make a real difference?*

The key differences between these two models are profound:

Feature	Traditional Public Administration	Results-Based Management (RBM)
Focus	Inputs (e.g., budget, personnel) and processes (e.g., following rules).	Results (outputs, outcomes, and impact) and strategic goals.
Accountability	Primarily procedural. Accountability is for following the rules and spending the budget.	Primarily for performance and results. Accountability is for achieving stated objectives.
Decision-Making	Based on precedent, rules, and hierarchical authority.	Based on evidence, data from monitoring, and feedback from evaluation.
Incentives	Reward compliance, seniority, and loyalty to the bureaucracy.	Reward performance, innovation, and achievement of results.
Role of M&E	Often a post-hoc, external requirement, if it exists at all.	An integral, continuous function for planning, tracking, learning, and reporting.

The transition to RBM is not merely a change in jargon; it requires a deep cultural transformation within the public sector. It demands a new way of thinking, where public servants at all levels see themselves not just as administrators of rules, but as managers of change and deliverers of value. This is the intellectual and operational space where M&E finds its purpose. The tools and frameworks of M&E—such as logical frameworks, performance indicators, and evaluation methodologies—are precisely what allow RBM to be implemented effectively.

1. The Role of Good Governance, Accountability, and Transparency

The institutionalization of Monitoring and Evaluation (M&E) in Africa's public sector is fundamentally a quest for good governance. M&E is not merely a technical tool for measuring performance; it is the engine that drives a virtuous cycle of accountability and transparency, without which genuine public sector reform is impossible. This section delves into the symbiotic relationship between M&E and these three core tenets of good governance, providing a comprehensive analysis of their importance and the challenges to their full realization.

1.3.1 The Pillars of Good Governance

Good governance is a broad concept, but for the purpose of our discussion, it can be broken down into three critical, interlocking components:

- **Accountability:** This is the obligation of public officials and institutions to answer for their actions, to report on their decisions and the use of public resources, and to accept responsibility for the outcomes. It operates in multiple directions: upward to political leaders and legislative bodies, downward to citizens and civil society, and horizontally among government institutions.
- **Transparency:** This refers to the free flow of information to the public. It means that government operations, policies, and decisions are open to scrutiny. Transparency builds public trust by making the inner workings of government visible and accessible, dispelling the myths and secrecy that often fuel corruption and inefficiency.
- **Participation:** This is the process through which citizens and civil society organizations are actively involved in the design, implementation, and oversight of public policies and programs. Participation ensures that governance is not a top-down exercise but a collaborative effort that is responsive to the needs and priorities of the people.

These three principles are not abstract ideals; they are practical requirements for an effective public sector. When they are absent, public administration tends to become insular, inefficient, and susceptible to corruption.

1.3.2 M&E as a Driver of Accountability

Monitoring and Evaluation systems are the very mechanisms through which accountability is made tangible and actionable. In a public sector that lacks institutionalized monitoring and evaluation (M&E), accountability is often a rhetorical concept. Officials may be held to account for following procedures, but not for achieving results. A functional M&E system changes this dynamic by:

- **Establishing a clear basis for accountability:** M&E frameworks define what constitutes success and failure through the use of specific, measurable indicators and targets. This moves accountability beyond vague promises and into the realm of evidence.
- **Providing the evidence base for oversight:** Parliaments, audit institutions, and other oversight bodies can use M&E reports and data to hold the executive branch accountable for its performance. Instead of relying on political rhetoric, they can demand concrete evidence of progress on national development plans.
- **Enabling citizen oversight:** When M&E data and findings are made public, citizens and civil society organizations can use this information to demand better performance from their government. This "bottom-up" accountability is a powerful check on the misuse of power and a driver for improved service delivery.
- **Fostering internal accountability:** Within a ministry or department, M&E data can be used to hold different units and managers responsible for their specific contributions to a program's success. This fosters a culture where performance is valued and where poor performance has consequences.

In the African context, where many countries have a history of weak oversight institutions, M&E offers a crucial tool for strengthening democratic checks and balances and re-establishing the social contract between the state and its citizens.

1.3.3 The Link Between M&E and Transparency

Transparency is the precondition for effective accountability, and M&E is a primary vehicle for achieving it. A government that is not transparent about its plans, budgets, and performance cannot be held accountable. Institutionalizing M&E promotes transparency by:

- **Making plans and budgets public:** A robust M&E system requires that national development plans, sectoral strategies, and program budgets are clearly articulated and made available to the public. This sets the stage for a transparent conversation about what the government intends to achieve and with what resources.
- **Publishing performance data and reports:** The results of monitoring and evaluation—including data on performance indicators, progress reports, and evaluation findings—must be regularly published in accessible formats. This allows citizens, the media, and researchers to independently verify government claims and identify areas of concern.
- **Demystifying government work:** M&E reports can translate complex government activities into understandable information about service delivery. For example, instead of just reporting on the number of schools built, M&E can report on the number of children in school and their learning outcomes, which is a more transparent measure of success.
- **Combating corruption:** Transparency in M&E can expose corruption by providing an auditable trail of where resources were allocated and what results were achieved. A discrepancy between reported expenditures and observable outcomes can be a red flag for malfeasance.

1.3.4 Challenges to Accountability and Transparency in Africa

Despite the clear benefits, institutionalizing M&E to promote good governance faces significant challenges in many African countries:

- **Political Resistance:** Political leaders and public officials who are used to operating with minimal scrutiny may resist the adoption of M&E systems that introduce transparency and accountability. The fear of being exposed for poor performance or corruption can be a powerful barrier.
- **Weak Institutional Capacity:** Many African countries lack strong, independent oversight institutions, such as a well-resourced and autonomous Auditor General's office, a vibrant and free press, and an empowered parliament, to effectively use M&E data.
- **Data Scarcity and Quality:** A lack of reliable and timely data can undermine M&E efforts. In some cases, data may be deliberately manipulated to show favourable results, a practice known as "data fudging," which is a direct antithesis to transparency.
- **Cultural Barriers:** In some contexts, a culture of deference to authority and a lack of a strong tradition of public scrutiny can hinder the demand for accountability from citizens.
- **Fragmented Systems:** M&E efforts are often fragmented across different ministries and projects, making it difficult to get a coherent, government-wide picture of performance and to hold the entire administration accountable.

1.3.5 Strategies for Strengthening Governance through M&E

Overcoming these challenges requires a concerted and multi-pronged approach:

- **Legislating for M&E:** Enacting laws that mandate M&E and the public disclosure of its findings can provide a strong legal basis for transparency and accountability. This makes M&E a legal requirement rather than a discretionary activity.

- **Empowering Oversight Institutions:** Providing more resources, training, and independence to parliaments, audit institutions, and anti-corruption bodies can equip them to effectively use M&E data for oversight.
- **Building a 'Demand Side' for M&E:** Governments must actively engage and empower civil society organizations, the media, and citizens to demand and use M&E information. This includes making data accessible, understandable, and widely publicized.
- **Leveraging Technology:** Digital platforms, open data portals, and mobile-based M&E systems can dramatically increase transparency by making data available in real-time and in user-friendly formats.
- **Championing from the Top:** The most successful cases of M&E institutionalization are often led by political champions who understand its value for good governance and are willing to push for it despite internal resistance.

Institutionalization of M&E in Africa's public sector is fundamentally about building a more effective, accountable, and transparent state. It is a reform agenda that places citizens at the centre, demanding that public resources are not just spent, but spent wisely, and that every public action is guided by the ultimate goal of improving people's lives.

1. **Case Study: The New Public Management (NPM) Movement and Its Influence on African States**

The evolution of public administration in Africa cannot be fully understood without a detailed examination of the New Public Management (NPM) movement. Emerging in the Anglo-Saxon world in the 1980s, NPM represented a radical departure from the traditional, Weberian bureaucratic model. Its influence spread rapidly across the globe, becoming a central tenet of public sector reform agendas, particularly in developing countries. This section serves as a comprehensive case study, analysing the origins of NPM, its core principles, its introduction to the African continent, and the profound, often contentious, impacts it has had on public administration and governance.

1. The Genesis and Core Principles of New Public Management

Born from the political philosophies of leaders like Margaret Thatcher in the UK and Ronald Reagan in the US, NPM was a direct response to the perceived inefficiencies, rigidities, and high costs of the post-war welfare state. Proponents of NPM argued that government bureaucracy was a major impediment to economic growth and that a fundamental re-engineering of the public sector was necessary.

The core tenets of NPM, as articulated by scholars like Christopher Hood, can be summarized as follows:

- **A "Hands-on" Management Approach:** A shift from career civil servants as "administrators" to a new breed of professional "managers" with greater autonomy and discretion to achieve results.
- **Explicit Performance Standards and Measures:** The introduction of clear, measurable targets and performance indicators to evaluate the success of programs and the performance of individual public servants.

- **Emphasis on Output Controls:** A fundamental shift in focus from the inputs (budgets, personnel) and processes (rules, procedures) of government to the outputs (services delivered) and outcomes (the impact on citizens' lives).
- **Disaggregation and Decentralization:** The breaking up of large, monolithic government departments into smaller, more autonomous executive agencies or units. This was intended to improve responsiveness and managerial accountability.
- **Introduction of Competition:** The use of market-type mechanisms within the public sector, such as contracting out services, public-private partnerships, and competitive tendering, to drive efficiency and lower costs.
- **Private Sector Management Styles:** The adoption of management techniques common in the corporate world, including performance-based pay, fixed-term contracts for senior managers, and a focus on "customer" satisfaction.
- **Lean and Streamlined Structures:** An emphasis on downsizing, "rightsizing," and cost-cutting to reduce the size and expenditure of the public sector.

In essence, NPM was driven by the idea that government should "steer rather than row"—meaning it should set policy and strategic direction, but leave the actual "rowing," or service delivery, to a more efficient mix of private companies, semi-autonomous agencies, and other non-state actors.

1. The Arrival of NPM in Africa: The Era of Structural Adjustment

The introduction of NPM to Africa was not a spontaneous, internally-driven process. Rather, it was largely propelled by external forces, particularly the international financial institutions (IFIs) like the World Bank and the International Monetary Fund (IMF). In the 1980s and 1990s, many African nations were facing severe economic crises, characterized by crushing debt, budget deficits, and stagnating growth.

To secure loans and debt relief, these countries were often required to implement Structural Adjustment Programmes (SAPs) as a condition of assistance. These programs, heavily influenced by neoliberal economic theories, included a public sector reform component that was the African manifestation of the NPM agenda. The reforms were presented as the solution to chronic public sector inefficiency, corruption, and a bloated civil service.

The specific NPM-style reforms introduced through SAPs included:

- **Civil Service Downsizing:** Widespread retrenchment programs aimed at reducing the size and cost of the public wage bill.
- **Privatization and Commercialization:** The sale of state-owned enterprises (SOEs) and the introduction of user fees for public services like water, healthcare, and education.
- **Decentralization:** The devolution of some administrative functions to local government levels or the creation of autonomous public agencies to manage specific services.
- **Performance Contracts:** The introduction of performance-based agreements for senior public officials and the creation of explicit performance targets for public institutions.

1.4.3 Case Studies of NPM Implementation in African States

The application of NPM principles across Africa was not uniform, and its outcomes varied significantly. Let's examine a few notable examples.

Ghana: Ghana's public sector reforms in the 1990s and 2000s are a classic example of NPM-driven change. The country faced significant challenges, including a lack of accountability, widespread corruption, and a bloated civil service. Reforms included the creation of semi-autonomous executive agencies, such as the Ghana Revenue Authority, to improve performance and revenue collection. The reforms also introduced performance management systems and sought to decentralize administrative functions to district assemblies. While these initiatives led to some improvements in service delivery in certain sectors, they were often hampered by a lack of political will, insufficient resources, and a civil service culture that was resistant to change.

South Africa: Following the end of apartheid in 1994, the new democratic government embarked on a massive public sector reform agenda. While the primary goal was to transform a racially segregated bureaucracy into a representative and service-oriented one, the reform framework, particularly the Growth, Employment and Redistribution (GEAR) strategy, incorporated many NPM elements. Key initiatives included the creation of a Senior Management Service (SMS) with performance-based contracts, a focus on performance management, and a push for greater decentralization. However, the implementation was inconsistent. For instance, the emphasis on performance management was often sporadic, and while decentralization was legislated, a lack of capacity and a re-strengthening of central state power in subsequent years limited its full realization.

Zambia and Zimbabwe: In countries like Zambia and Zimbabwe, NPM-style reforms were evident in the push for decentralization and the introduction of autonomous agencies. In the 1990s, for example, both countries experimented with the "hiving-off" of services, such as the creation of autonomous hospitals and the contracting out of certain services like road maintenance. While some studies showed cost savings and improved efficiency in these specific areas, the reforms often failed to address the systemic issues of political instability and broader governance deficits, ultimately limiting their long-term impact.

1. The Critiques and Mixed Legacy of NPM in Africa

The implementation of NPM in Africa has been a subject of intense debate. While proponents point to its role in modernizing public administration and introducing a focus on results, critics argue that it has had a mixed, and in many cases, detrimental impact. The main criticisms include:

- **The "One-Size-Fits-All" Fallacy:** NPM, a model developed for the specific conditions of Western, industrialized nations, was often applied to African countries without sufficient consideration of their unique cultural, institutional, and political contexts. This led to a disjuncture between the reform's design and its implementation on the ground.
- **Erosion of State Capacity:** The push for downsizing and fiscal austerity, a core component of many SAPs, often led to the retrenchment of skilled and experienced public servants. This exodus of talent, particularly in key technical ministries, weakened the state's capacity to formulate and implement its own policies, making it more reliant on external consultants.
- **Weakening of Public Service Ethos:** The introduction of private sector values, such as "customer" focus and financial incentives, was criticized for potentially eroding the traditional ethos of public service, which is built on principles of equity, social justice, and impartiality. The focus on measurable outputs could also lead to a neglect of less-

measurable, but equally important, public goods.

- **The Problem of Data Fudging:** In a context where performance contracts and explicit targets are linked to rewards or sanctions, there is a strong incentive for public officials to "game the system" by manipulating data or focusing only on what is easy to measure, rather than what is genuinely important. This undermines the very purpose of M&E and accountability.
- **Failure to Address Governance Deficits:** NPM often focused on technical and managerial fixes while sidestepping the deeper political and governance issues, such as corruption, a lack of the rule of law, and weak political will. In many cases, these underlying issues acted as binding constraints, preventing the reforms from achieving their intended results.

1.4.5 The Evolution to "Post-NPM" and the Future of Public Administration in Africa

The critiques of NPM have led to a recognition that a purely market-oriented model is insufficient for the complexities of public service. This has given rise to a "post-NPM" paradigm, which seeks to re-emphasize the importance of the public good, democratic accountability, and a capable, professional civil service. This new approach recognizes that while efficiency is important, it must be balanced with equity, participation, and social justice.

In the African context, this evolution is evident in the shift from a passive, externally-driven reform agenda to one that emphasizes national ownership, building internal capacity, and strengthening M&E systems as a tool for learning and adaptation, rather than just as a mechanism for external accountability. The focus is no longer just on "how to manage better," but on "how to govern better."

This extensive case study provides a critical backdrop for our book. It illustrates that while the principles of NPM introduced a valuable focus on results, their uncritical application in Africa led to significant challenges. The lesson for today is that for M&E to be institutionalized effectively, it must be part of a broader, nationally-owned reform agenda that is sensitive to local contexts and committed to a holistic vision of good governance.

Chapter 2

Understanding Monitoring and Evaluation (M&E)

Introduction

The shift in public sector management from a focus on process to one on results is underpinned by a deep understanding and application of Monitoring and Evaluation (M&E). While often used as a single phrase, "M&E" represents two distinct but interconnected functions. To institutionalize M&E effectively, policymakers and practitioners must first have a clear and comprehensive grasp of what monitoring is, what evaluation is, and how they work together to drive a culture of evidence-based decision-making. This chapter is dedicated to providing that foundational understanding, moving beyond simple definitions to explore the purpose, methodologies, and practical application of M&E in the African public sector context.

2.1 Defining M&E: What is Monitoring vs. Evaluation?

A common misconception is that monitoring and evaluation are the same thing, or that they are simply about data collection and report writing. This is a crucial misunderstanding that can lead to ineffective programs and a failure to learn from experience. In reality, they are two separate but symbiotic functions, each with a unique purpose, timing, and set of questions it seeks to answer.

2.1.1 What is Monitoring?

Monitoring is a continuous and systematic process of collecting, analysing, and using data on the indicators of a project, program, or policy. Its primary purpose is to provide management and stakeholders with regular feedback on the progress of an intervention. Think of it as the "eyes and ears" of a program, constantly checking its pulse to ensure it is healthy and on track.

Key Characteristics of Monitoring:

- **Timing:** Monitoring is a continuous activity that takes place throughout the life cycle of a program, from the start of its implementation to its end. It is a real-time, ongoing process.

- **Purpose:** The main objective is to track progress. It provides an early warning system, allowing managers to identify problems and bottlenecks as they arise. It answers questions like: *Are we doing what we said we would do? Are we on schedule? Are we reaching our target beneficiaries? Are we within our budget?*
- **Focus:** The focus of monitoring is primarily on inputs, activities, and outputs.
    - **Inputs:** The resources used for the program (e.g., money, personnel, equipment). Monitoring tracks if these inputs are being delivered as planned.
    - **Activities:** The actions taken to achieve the program's objectives (e.g., training workshops conducted, vaccines administered, roads constructed). Monitoring checks if these activities are happening.
    - **Outputs:** The direct products of the activities (e.g., number of people trained, number of children vaccinated, kilometres of road built). Monitoring quantifies these direct results.
- **Who does it?** Monitoring is typically an internal function, carried out by the program implementers themselves or a dedicated M&E unit within the implementing organization. This is because it is a management tool for internal decision-making.

Example in an African Public Sector Context: Consider a national government program to build 1,000 new classrooms across the country. Monitoring would involve a dedicated team systematically tracking:

- The number of contracts awarded to builders (Input).
- The number of construction sites active and the progress on each (Activity).

- The number of classrooms completed and handed over (Output).
- The expenditure every month against the budget (Input).

The monitoring report would inform the Ministry of Education if the projects were behind schedule in a particular region or if a contractor is not meeting their targets, allowing for timely intervention.

2.1.2 What is Evaluation?

Evaluation, in contrast to monitoring, is a more rigorous and periodic assessment of an intervention. It is a systematic and objective process to determine the relevance, effectiveness, efficiency, impact, and sustainability of a program or policy. If monitoring is the "eyes and ears" that watch the program as it runs, evaluation is the "brain" that analyses its performance at key milestones and the end of its life. It seeks to answer the crucial question of *why* something worked or didn't, and what difference it made.

Key Characteristics of Evaluation:

- **Timing:** Evaluation is a periodic activity, often conducted at specific points in a program's life cycle (e.g., mid-term, end-of-project) or as a post-implementation assessment. It is a retrospective process, looking back to draw lessons.
- **Purpose:** The goal is to make a judgment about a program's worth, to understand its successes and failures, and to inform future policy and strategy. It answers broader questions than monitoring, such as: *Was this program the right solution? Did it achieve its intended outcomes? Was it the most cost-effective way to achieve those results? What were the unintended impacts?*
- **Focus:** Evaluation focuses on outcomes and impact.
    - **Outcomes:** The short-term and medium-term effects of a program's outputs (e.g., an increase in the number of children

enrolled in school as a result of the new classrooms).

- **Impact:** The long-term, lasting changes brought about by the program (e.g., improved literacy rates in the country, a more skilled workforce in the long run).

- **Who does it?** Evaluations are often carried out by external, independent experts or a dedicated unit with a degree of autonomy. This is to ensure objectivity and impartiality in the assessment.

Example in an African Public Sector Context: Following the completion of the 1,000 classrooms project, an evaluation would be conducted to answer questions such as:

- Did the new classrooms lead to a significant increase in primary school enrolment? (Outcome)
- Did the quality of education improve in the new schools? (Impact)
- Was building the classrooms the most cost-effective way to improve education, or would a different intervention (e.g., teacher training) have had a greater impact? (Efficiency & Relevance)
- Are the communities maintaining the classrooms without government support? (Sustainability)

2.1.3 The Interplay: A Continuous Feedback Loop

The true power of M&E lies in their collaboration. They are not isolated activities but two parts of a continuous feedback loop that drives organizational learning and improvement.

Aspect	Monitoring	Evaluation
Question	Are we doing what we planned?	Is what we are doing working?
Timing	Continuous, ongoing	Periodic, retrospective
Focus	Inputs, activities, outputs	Outcomes, impact, relevance
Primary Use	Course correction, daily management	Strategic decisions, future planning
Staffing	Internal program staff	External, independent experts

Monitoring data serves as the foundation for evaluation. An evaluator cannot assess a program's impact without knowing what activities were actually carried out and what outputs were delivered, all of which is provided by the monitoring system. Conversely, the findings of an evaluation—about what worked and why—inform future program design and provide new indicators for monitoring.

That's an excellent place to start. A clear and comprehensive definition of M&E is the essential foundation for the entire book. I will provide a detailed, five-page discussion on this topic, breaking it down to ensure every aspect is covered.

2.2 The Purpose and Value of M&E in the Public Sector

The institutionalization of Monitoring and Evaluation (M&E) is not an end in itself; it is a means to a greater end: the improvement of public sector performance and, ultimately, the delivery of better results for citizens. In the context of Africa's evolving governance landscape, M&E serves three fundamental and interconnected purposes. It is a tool for strategic learning, a mechanism for strengthening accountability, and a powerful instrument for evidence-based resource allocation. Together, these functions are the primary drivers of M&E's value and the reason why its institutionalization is a non-negotiable component of modern public administration.

2.2.1 M&E as a Tool for Learning and Improvement

One of the most profound values of a well-functioning M&E system is its capacity to transform a government into a "learning organization." In traditional public administration, a program's life cycle often ends with its completion, and its successes and failures are rarely systematically documented or used to inform future interventions. M&E breaks this cycle by creating a continuous feedback loop that fosters institutional learning.

- **Understanding What Works and Why:** Monitoring provides real-time information on whether a program is on track, but it is evaluation that provides the deeper insights. An evaluation can reveal that a health program, for example, failed to reduce a specific disease, not due to a lack of funding, but because its design was flawed, or because the cultural context of the target community was misunderstood. This evidence allows policymakers to understand the causal links between their actions and the outcomes, helping them to avoid repeating past mistakes and to replicate successful models.
- **Encouraging Policy Redesign:** In countries like Uganda, M&E has been used to assess the performance of projects, such as those under the Global Partnership for Education. These assessments help to identify challenges in implementation and inform the redesign of subsequent projects. For example, if monitoring data consistently shows that a water project is not meeting its targets, an evaluation may be commissioned to discover that the technology used is not suitable for the local environment. This learning then leads to a redesign of the project, ensuring that the next intervention is more effective and appropriate.
- **Fostering a Culture of Reflection:** A truly institutionalized M&E system fosters a culture where program managers, civil servants, and policymakers are encouraged to reflect on their performance, ask critical questions, and seek evidence-based solutions. This intellectual curiosity, driven by M&E findings,

is the engine of innovation and continuous improvement in the public sector.

2.2.2 M&E as a Mechanism for Accountability

Accountability is a cornerstone of good governance, and M&E is its most powerful technical enabler. By generating credible information on performance and results, M&E makes it possible to hold government institutions, officials, and political leaders accountable for their actions and for the use of public funds. This accountability operates on two critical levels:

- **Upward Accountability to Political and Legislative Bodies:** M&E provides a vital link between the executive branch and legislative oversight bodies, such as parliaments. When governments are required to report on their progress toward national development goals, they use M&E data to show what they have achieved. This information empowers parliamentary portfolio committees and other oversight bodies to scrutinize government performance, ask difficult questions, and hold ministers accountable for the delivery of their mandates. For example, in South Africa, the Department of Planning, Monitoring, and Evaluation (DPME) submits reports to Parliament on the performance of national programs, which is a core function of the country's accountability framework.
- **Downward Accountability to Citizens and Civil Society:** In a democratic context, accountability also flows downwards to the public. Making M&E reports and data publicly available through websites, dashboards, and other portals empowers citizens, civil society organizations, and the media to monitor government performance. This transparency strengthens public trust, reduces the opportunities for

corruption, and enables a more informed public debate on government policies. The Twende Mbele network, which promotes learning between countries like Kenya, Benin, and Uganda, emphasizes the importance of using M&E to enhance transparency toward citizens. By providing credible evidence of results, governments can build trust and demonstrate their commitment to serving the needs of the people.

2.2.3 M&E as an Instrument for Evidence-Based Resource Allocation

In a context of limited resources, one of the most significant values of M&E is its ability to inform smarter, more effective resource allocation. Traditional budgeting often relies on historical precedent, political considerations, or simply incremental increases. M&E allows governments to move beyond this to a more strategic, performance-based approach.

- **Linking Budgeting to Results:** M&E provides the necessary evidence to implement performance-based budgeting. This means that instead of just funding ministries based on what they spent last year, a government can use M&E findings to allocate resources based on what a program has actually achieved. For instance, an M&E report might show that a specific road maintenance program in a particular region is more cost-effective and generates better economic outcomes than a similar program in another region. Based on this evidence, the Ministry of Finance can reallocate funds to the more successful program, thereby maximizing the return on public investment. The Twende Mbele network has published guidelines and case studies from countries like South Africa and Benin that detail how M&E evidence is used to inform budget and planning processes.

- **Identifying Inefficiencies and Waste:** Regular monitoring can quickly identify programs that are over budgeted, behind schedule, or failing to produce their intended outputs. This provides an early warning system that allows managers to make timely course corrections. Similarly, evaluations can pinpoint systemic inefficiencies or interventions that are simply not working, providing the evidence needed to terminate ineffective programs and re-purpose their funding toward more promising alternatives. This not only prevents wasteful spending but also signals to the public that the government is a responsible steward of taxpayer money.

The purpose and value of M&E are inextricably linked. It provides the evidence needed for a government to learn from its experiences, the information required to hold itself accountable to its citizens, and the data necessary to ensure that public resources are allocated in the most effective and efficient manner possible. For African public sector management, M&E is not a luxury; it is a foundational tool for building a more responsive, accountable, and results-oriented state.

2.3 Key M&E Concepts: The Building Blocks of a Results-Based System

The effective institutionalization of M&E requires a deep understanding of its foundational concepts. These are the building blocks that allow a government to move from abstract intentions to measurable results. They provide the logical structure and the technical precision needed to track progress, assess performance, and make evidence-based decisions. This section provides a comprehensive discussion of the four key concepts that form the basis of any M&E system: the results chain, indicators, baselines, and targets.

2.3.1 The Results Chain: Mapping the Path to Change

At the heart of every effective M&E system is a results chain, often referred to as a Theory of Change or a logical framework. This is a logical diagram or narrative that maps out the causal links between a program's activities and its intended impact. It moves beyond a simple statement of intent and forces program designers to articulate their assumptions about how change will actually happen. A typical results chain consists of a logical flow from Inputs to Impact:

- **Inputs:** These are the resources mobilized to carry out the program or project. They include financial resources, human capital (staff), physical assets (e.g., vehicles, buildings), and equipment. Without adequate inputs, no activities can take place.
- **Activities:** These are the actions taken using the inputs. Examples include conducting training workshops, constructing a school building, or administering vaccines to children. These are the direct actions of the implementing organization.
- **Outputs:** These are the direct, tangible products or services that result from the activities. They are typically quantifiable and under the direct control of the program or project. For example, a training program's output is the *number of people trained* or the *number of training modules developed.* A road construction program's output is the *kilometres of road built.*
- **Outcomes:** These are the short- to medium-term effects or changes that occur as a result of the outputs, whether negative or positive. They are the changes in knowledge, attitudes, skills, or behaviours of the target beneficiaries. For a training program, an outcome might be a *10% increase in the knowledge of trainees* or *the adoption of new agricultural techniques by smallholder farmers*. Outcomes are the "so what"

of the outputs.

- **Impact:** This is the ultimate, long-term, and often large-scale change to which the program contributes. It is the final development result. For a training program aimed at farmers, the impact might be an *increase in national food security* or an *improvement in rural livelihoods*. Impact is rarely the result of a single program and is often influenced by many external factors.

Example: A National Water and Sanitation Program
Let's illustrate with a hypothetical program in the public sector:

- **Inputs:** $10 million in government and donor funds, 20 engineers and technical staff.
- **Activities:** Drill 50 new boreholes, hold hygiene workshops for 100 communities.
- **Outputs:** 50 new boreholes installed, 100 communities participated in workshops.
- **Outcomes:** 80% of targeted households report having access to a new water source within a 15-minute walk. The incidence of waterborne diseases in the targeted communities decreases by 50%.
- **Impact:** An overall improvement in public health and a reduction in poverty due to fewer sick days and a healthier workforce.

A well-defined results chain makes it possible to track the logical progression of change and to identify where the causal link may be broken.

2.3.2 Indicators: Measuring What Matters

An **indicator** is a specific, measurable variable that provides a simple and reliable means to measure progress toward achieving a result. Indicators are what turn the qualitative statements in a results chain into a basis for a rigorous M&E system. Without indicators, a results chain remains a wish list rather than a roadmap for action.

Good indicators are often defined by the **SMART** criteria:

- **Specific:** They must be clear and unambiguous.
- **Measurable:** They must be quantifiable or qualifiable in a consistent manner.
- **Achievable:** The data for the indicator must be possible to collect.
- **Relevant:** They must directly measure the intended result.
- **Time-bound:** They should be tied to a specific timeframe.

Indicators are required for every level of the results chain:

- **Output Indicators:** *Number of boreholes installed. Number of communities trained.*
- **Outcome Indicators:** *Percentage of households with access to safe drinking water. Percentage decrease in waterborne diseases.*
- **Impact Indicators:** *Percentage change in the regional poverty rate. Life expectancy at birth in the targeted region.*

The South African Legislative Sector, for instance, uses a database of some 500 performance indicators to track government performance against presidential goals, demonstrating how indicators are central to a national accountability framework.

2.3.3 Baselines: Establishing the Starting Line

A baseline is the value of an indicator at the very beginning of a program or policy intervention, before any activities have commenced. It is the starting point, the "before" picture against which all future progress is measured.

The importance of a baseline cannot be overstated. Without it, it is impossible to:

- **Measure Change:** You cannot determine if a program has made a positive, negative, or no change at all. For example, if a health program aims to reduce a specific disease, the baseline must establish the prevalence of that disease before the program starts.
- **Attribute Results:** A baseline helps to attribute observed changes to a program. While it is challenging, a robust baseline is a critical part of a rigorous evaluation that helps to answer the question, "Would this change have happened anyway without our intervention?"
- **Set Realistic Targets:** Realistic targets must be informed by a program's starting point. A government cannot set an ambitious target of a 50% increase in school enrolment without knowing the baseline enrolment rate.

In many African countries, the lack of reliable pre-existing data is a major challenge for establishing baselines. Best practice often requires a specific baseline study to be conducted as part of the project design phase to collect this crucial information. The Government of Zimbabwe, in its National Monitoring and Evaluation Policy, underscores the importance of baselines by mentioning the use of census data to provide starting-point estimates for key national indicators, such as primary education enrolment rates.

2.3.4 Targets: Setting the Destination

A target is the specific, time-bound, and quantitative goal that a program aims to achieve for a given indicator. Targets transform an M&E system from a simple data collection exercise into a forward-looking tool for performance management. They define the expected level of achievement by a specific date.

Targets are crucial for:

- **Motivation and Direction:** They provide a clear destination for program managers and staff, motivating them to work towards a common goal.
- **Accountability:** They establish clear expectations against which performance can be judged. When a program fails to meet a target, it triggers an investigation to understand why, as is the practice in the South African public sector, where managers must provide a statement explaining their underperformance.
- **Resource Management:** They inform strategic decision-making. If a target is not being met, it may signal that more resources are needed or that a different approach is required.

Targets must be ambitious enough to drive performance but also realistic enough to be achievable. They should be set in a participatory manner, involving all key stakeholders to ensure ownership and buy-in. A national policy might set a long-term target (e.g., "reduce the national poverty rate by 15% in 10 years"), while a project would have shorter-term targets (e.g., "train 500 health workers by the end of the first year").

2.3.5 The Interplay: An Integrated Framework

These four concepts are not isolated; they form a logical, integrated framework. The results chain is the strategic roadmap that outlines the causal logic of the intervention. The indicators are the metrics used to measure progress along that map. The baseline is the starting point from which all change is measured. And the targets are the milestones and the ultimate destination on the journey.

A failure in one component can cripple the entire system. For instance, without a clear results chain, you cannot select the right indicators. Without a baseline, you cannot set realistic targets. And without both, M&E becomes a meaningless exercise in data collection without any means to judge the program's worth. The effective institutionalization of M&E requires all four of these concepts to be understood, developed, and applied in a coordinated and systematic manner.

2.4 Types of Evaluation: A Strategic Toolbox for Government

Evaluation is not a monolithic concept; it is a strategic toolbox containing different instruments, each designed to answer a specific set of questions at a particular point in a program's life cycle. The purpose of an evaluation—whether to improve a program, make a final judgment, or understand its long-term effects—dictates the type of evaluation to be conducted. A truly institutionalized M&E system understands this and deploys the right evaluation type at the right time. This section provides a comprehensive overview of the key types of evaluation and their strategic use in the African public sector.

2.4.1 Formative Evaluation (and Process Evaluation)

A **formative evaluation** is a backward-looking assessment conducted during the early or mid-stages of a program to provide feedback on its design and implementation. Its primary purpose is not to make a final judgment of worth, but rather to "form" and improve the program while it is still running.

- **Purpose:** The key questions a formative evaluation seeks to answer are: *Is the program being implemented as planned? Are there any unexpected challenges or bottlenecks? Is the program's theory of change still relevant?* It is an iterative, learning-focused exercise designed to help managers make mid-course corrections and optimize the program's effectiveness.
- **Timing:** Formative evaluations are conducted early in a program's life, typically within the first one or two years of implementation.
- **Relevance to Africa:** In resource-constrained environments,

formative evaluations are invaluable. They prevent a government from continuing with a flawed program and wasting valuable public funds. For instance, a formative evaluation of a new national health clinic program could be used to assess if the clinics are being built in the right locations, if the medical equipment is being delivered as planned, and if the local communities are using the services. The findings could reveal that a particular clinic is being underutilized due to a lack of public transport, leading the government to make a strategic adjustment to its location or to provide a transport subsidy.

Process Evaluation: A **process evaluation** is a specific type of formative evaluation that focuses on how a program is being implemented. It is a detailed assessment of a program's internal workings. A process evaluation answers the question: *Are the program's activities being carried out as intended and with the fidelity required to produce the desired outputs?* This type of evaluation is a crucial precursor to a summative or impact evaluation, as it provides the necessary evidence to confirm that the program was actually delivered as designed.

- **Example from the African Public Sector:** In a project to improve literacy rates in schools, a process evaluation could be conducted to check if the new curriculum is being taught by teachers as planned, if the school libraries are receiving the books they were allocated, and if the children are attending classes regularly. A World Bank-led evaluation of an education project in Uganda used process evaluation to assess the delivery of program activities, identify where implementation was straying from the original design, and inform program managers on how to get it back on track.

2.4.2 Summative Evaluation (and Impact and Outcome Evaluation)

A **summative evaluation** is a backward-looking assessment conducted after a program to determine its overall worth, success, or failure. Its purpose is to make a final judgment on a program's effectiveness.

- **Purpose:** The key questions a summative evaluation seeks to answer are: *Did the program achieve its objectives? Was it worth the investment? Should this program be continued, replicated, or terminated?* Summative evaluations are often conducted to meet accountability requirements and to inform strategic decisions about future policy and resource allocation.
- **Timing:** Summative evaluations are typically conducted at the end of a program's life or key strategic milestones.
- **Relevance to Africa:** Summative evaluations are vital for accountability in a context where public funds are scarce and public trust is fragile. By providing a final, evidence-based judgment of a program's success, they allow governments to demonstrate to their citizens and to development partners that public money has been well spent.

Impact Evaluation: An **impact evaluation** is the most rigorous and sophisticated form of summative evaluation. Its purpose is to answer the most difficult question in M&E: *What is the causal impact of the program?* This requires establishing a "counterfactual"—what would have happened to the beneficiaries if they had not participated in the program. Without a counterfactual, it is impossible to know if observed changes were due to the program or to other external factors.

- **Methodology:** Impact evaluations use a variety of robust methodologies to establish a counterfactual, including:
    - **Randomized Controlled Trials (RCTs):** Considered the "gold standard," this involves randomly assigning individuals or

groups to either a treatment group (that receives the program) or a control group (that does not). Any difference in outcomes between the two groups can then be attributed to the program.
   - **Quasi-Experimental Designs:** These methods are used when randomization is not possible. They include techniques like propensity score matching and difference-in-differences, which use statistical analysis to create a comparable control group.
- **Example from the African Public Sector:** The International Initiative for Impact Evaluation (3ie) and other research organizations have conducted numerous impact evaluations in Africa. A classic example is the evaluation of a malaria prevention program in a sub-Saharan African country. Using an RCT, researchers provided mosquito nets to a randomly selected group of villages (the treatment group) and did not provide them to another, equally selected group (the control group). The evaluation found that the program significantly reduced malaria prevalence, providing the evidence needed for the government to scale up the program.

Outcome Evaluation: An **outcome evaluation** is a less methodologically rigorous form of summative evaluation that focuses on a program's short- to medium-term results. It answers the question: *Did the program achieve its intended outcomes?* While it may not be able to definitively establish a causal link, it can provide valuable evidence on whether the program is producing the desired results.

2.4.3 A Strategic, Interconnected Approach to Evaluation

The different types of evaluation are not mutually exclusive; a robust M&E system uses them in a coordinated and strategic manner throughout a program's life cycle. This is a journey from initial design to final judgment, each stage building on the last.

- **At the Beginning:** A **formative evaluation** is conducted to ensure the program design is sound and its implementation is on the right track.

- **During Implementation:** A **process evaluation** is conducted continuously or periodically to ensure the program is being delivered with the fidelity and quality needed to achieve its objectives.
- **At the End:** A **summative evaluation** is conducted to make a final judgment of worth, often consisting of both an **outcome evaluation** to determine if the short-term results were achieved and a more rigorous **impact evaluation** to definitively determine the program's causal effects.

The following table summarizes the key differences:

Evaluation Type	Key Question	Timing	Primary Purpose	Methodology
Formative	Is the program design sound?	Early in program life	To inform and improve	Qualitative, site visits
Process	Is the program being implemented as planned?	Throughout implementation	To check fidelity	Observation, interviews, log reviews
Outcome	Did the program achieve its short-term results?	End of program	To make a judgment of merit	Performance data analysis, surveys
Impact	What is the causal effect of the program?	End or well after program	To attribute change to the program	RCTs, quasi-experimental designs

Understanding the different types of evaluation is a prerequisite for effective M&E. By strategically deploying the right tool for the right job, African governments can move beyond simple reporting to a system of rigorous inquiry that drives a deeper understanding of what works, what doesn't, and why—thereby ensuring that every public investment contributes to the long-term development of the continent.

Chapter 3

The Concept of Institutionalization

Introduction

Institutionalization of Monitoring and Evaluation (M&E) is the comprehensive, multi-dimensional process of embedding M&E into the core functions, structures, and culture of a government or public sector organization. It transcends the temporary, project-based practice of "doing" M&E to become a permanent, self-sustaining system. This involves establishing M&E with a legal and policy mandate, integrating its processes into the national planning and budgeting cycles, building a permanent cadre of skilled M&E professionals, and, most importantly, fostering a political and cultural environment where M&E evidence is routinely demanded and used for decision-making, accountability, and continuous organizational learning. Ultimately, institutionalization transforms M&E from an external requirement into an internal driver of good governance and effective public management.

3.2 Institutionalization: The Journey from Ad-Hoc Activity to Core Government Function

The concept of Monitoring and Evaluation has gained widespread currency in Africa's public sector, but its implementation remains highly uneven. On one end of the spectrum, M&E is treated as a temporary, ad-hoc activity; on the other, it is a deeply embedded, systemic function of government. The distinction between these two approaches is not just a matter of scale—it represents a fundamental difference in philosophy, purpose, and impact. This section explores why the institutionalization of M&E is different from simply "doing" it and why this journey is essential for sustainable development.

3.2.1 The "Doing" M&E Approach: A Tale of Projects

In many parts of Africa, the practice of M&E has historically been a reactive, rather than a proactive, endeavour. It has been driven largely by external pressures, particularly the reporting requirements of development partners and international donors. This approach can be characterized as "doing" M&E—a series of project-specific, often disconnected activities that are not integrated into the broader machinery of government.

Key characteristics of this ad-hoc approach include:

- **Donor-Driven Mandates:** M&E is often a mandatory component of a donor-funded project. A logframe is created, a project-level M&E team is hired on a temporary contract, and reports are generated to satisfy the donor's accountability requirements.
- **Project-Specific and Unsustainable:** The M&E system is designed and built for a single project. When the project ends, the M&E team is disbanded, and the system—along with the data, lessons learned, and capacity built—often disappears. There is no transfer of knowledge to the next project or the wider government.
- **Lack of Internal Demand:** The demand for M&E information comes from external stakeholders, not from the government itself. Public officials may view M&E as a bureaucratic chore, a necessary evil for securing funds, rather than a valuable tool for improving their own performance.
- **Findings Gather Dust:** In this environment, M&E findings are often treated as an endpoint rather than a starting point for action. Evaluation reports may be filed away, and their recommendations are rarely integrated into policy or used to inform budget allocation. The crucial feedback loop is broken.
- **Capacity is Temporary:** The capacity built is often limited to a few individuals hired for the project's duration. There is no strategic effort to build a permanent cadre of M&E professionals within the civil service.

In this model, M&E remains on the periphery of government operations. It is a technical exercise performed to tick a box, not a strategic function that guides public administration.

3.2.2 The Institutionalization of M&E: Embedding into the DNA of Government

Institutionalizing M&E is a far more profound and transformative process. It is about embedding the principles, practices, and structures of M&E into the very fabric of public administration. It is a transition from M&E as an event to M&E as a core, continuous function.

This journey is built on several key pillars:

- **Legal and Policy Frameworks:** Institutionalization begins with a clear legal and policy basis. This includes national M&E policies, strategies, and even legislation that mandates all government ministries and departments to conduct M&E, report on their results, and use the findings for planning and budgeting. This makes M&E a legal requirement, not a discretionary option.
- **Dedicated Structures and Mandates:** This involves establishing permanent M&E units or directorates within key government institutions, such as the Presidency, the Ministry of Finance, and line ministries. These units are staffed by career civil servants with clear terms of reference and mandates to oversee and coordinate M&E activities.
- **Budgetary Allocation:** A critical sign of institutionalization is the allocation of a dedicated budget line for M&E in national and sectoral budgets. This ensures that M&E is adequately resourced and no longer reliant on project-specific or donor-earmarked funds. Typically, a small percentage of a program's budget (e.g., 1-3%) is ring-fenced for M&E.
- **Human and Technical Capacity:** Institutionalization requires building a professional cadre of M&E experts within

the civil service. This involves long-term training, career paths, and professional development programs. It also entails investing in the necessary technical infrastructure, such as national data systems and M&E information portals.

- **Culture of Use and Learning:** Perhaps the most difficult pillar to build is a culture where M&E findings are systematically used for decision-making. This means integrating M&E data into policy formulation, annual budget processes, and parliamentary oversight. It transforms M&E from a simple reporting exercise into a powerful tool for learning and accountability.

3.2.3 A Comparative Perspective: Doing vs. Institutionalizing

To make this distinction clear, let us compare the two approaches across key dimensions:

Dimension	"Doing" M&E (Ad-Hoc Approach)	Institutionalizing M&E (Systemic Approach)
Purpose	To meet external reporting requirements.	To improve government performance and accountability.
Driver	External donors and project managers.	Internal demand from policymakers and citizens.
Scope	Project-specific, siloed.	National, government-wide, and integrated.
Funding	Project budgets, often donor-funded.	Dedicated line item in national and sectoral budgets.
Capacity	Temporary, project-based staff.	Permanent, professional civil service cadre.
Use of Findings	Often ignored or used for reports only.	Systematically integrated into planning, budgeting, and policy.
Sustainability	Ends when the project ends.	Permanent and self-sustaining over time.

3.2.4 The Imperative of Institutionalization in Africa

The challenges facing Africa's public sector—from resource scarcity and corruption to weak service delivery—cannot be solved by a piecemeal, project-based approach to M&E. The institutionalization of M&E is a strategic imperative because it:

1. **Ensures Sustainability:** It creates a permanent system that endures beyond the lifespan of any single project or government administration.
2. **Fosters a Learning Culture:** It provides a mechanism for continuous learning across government, allowing ministries to build on successes and avoid repeating past mistakes.
3. **Strengthens Good Governance:** By embedding M&E into the core functions of government, it strengthens accountability and transparency, thereby building public trust and legitimacy.
4. **Increases Efficiency:** It ensures that public resources are not just spent, but spent wisely, and that they contribute to achieving national development goals.

The journey from "doing" M&E to institutionalizing it is the journey from a passive, externally-driven activity to an active, nationally-owned tool for development. It is a journey that, while difficult and complex, is non-negotiable for any African state committed to effective, results-based public administration.

3.3 The Hallmarks of a Truly Institutionalized M&E System

Moving beyond the ad-hoc practice of M&E to its full institutionalization is a journey of structural and cultural transformation. A truly institutionalized M&E system is not a standalone unit or a one-off report; it is a complex, interconnected web of processes, policies, and people that are deeply embedded in the governance of a state. This section outlines the key characteristics and hallmarks of such a system, providing a blueprint for what a functional, results-oriented public sector looks like in practice.

3.3.1 A Foundation in Legal and Policy Frameworks

The first and most critical characteristic of an institutionalized M&E system is that it has a clear and explicit legal and policy foundation. Unlike a project-based M&E activity that is dependent on a funding agreement, an institutionalized system is a legal or policy requirement of the state itself.

- **Legal Mandates:** This can take the form of specific M&E laws, national constitutions, or legislative acts that mandate all government ministries and departments to conduct M&E. This provides the system with legitimacy and authority, making it a non-negotiable function of government.
- **National M&E Policies:** Many African countries have moved to develop dedicated national M&E policies. These policies provide an overarching framework, defining the principles, standards, roles, and responsibilities for M&E across all levels of government. Examples of countries that have adopted such policies include **South Africa**, **Uganda**, **Benin**, **Zimbabwe,** and **Zambia**. These policies make M&E a structured, government-wide endeavour rather than a series of isolated initiatives.
- **Strategic Alignment:** An institutionalized system's legal and policy frameworks ensure that M&E efforts are directly aligned with national development plans and strategies, such as a country's Vision 2030 or a five-year Medium-Term Expenditure Framework. This alignment guarantees that M&E resources are focused on tracking progress toward the nation's highest-level priorities.

3.3.2 Integration into Public Financial Management (PFM) Cycles

The ultimate value of M&E is realized when its findings are used to make better decisions about the allocation of scarce public resources. A truly institutionalized system seamlessly integrates M&E into the core processes of public financial management, specifically planning, budgeting, and oversight.

- **Evidence-Informed Planning and Budgeting:** In an institutionalized system, M&E data and evaluation evidence are not an afterthought; they are a fundamental input into the annual budget cycle. For instance, in South Africa, the Department of Planning, Monitoring, and Evaluation (DPME) has issued guidelines on how to use M&E evidence to inform strategic and annual performance plans. Ministries are required to justify their budget requests based on the performance of their past programs, as revealed by monitoring data and evaluation findings.
- **A Dedicated M&E Budget:** A telltale sign of institutionalization is the presence of a dedicated budget line for M&E activities. This ensures that M&E is seen as a necessary cost of doing business, rather than a cost that can be easily cut or ignored. This ring-fencing of funds for M&E, often a small percentage of a program's budget, provides the financial sustainability needed for the system to function effectively.
- **Performance-Informed Oversight:** Legislators and parliamentary oversight committees are key users of M&E information. A well-integrated system provides parliament with regular, standardized performance reports that allow it to hold the executive branch accountable for results, making budget approvals and policy debates more evidence-based.

3.3.3 A Robust and Coordinated Institutional Structure

The institutionalization of M&E requires a clear and effective organizational structure. This structure is typically a "whole-of-government" approach, comprising both a central coordinating body and dedicated M&E units within ministries.

- **Central M&E Body:** Many successful systems are anchored by a high-level central M&E unit, often located in a powerful ministry such as the Ministry of Finance, Planning, or even the Presidency. This strategic location gives the unit the political authority to set standards, coordinate activities, and hold line ministries accountable. The M&E systems in Uganda and South Africa, located in the Office of the Prime Minister and the Presidency, respectively, serve as powerful examples of this.

- **Decentralized M&E Units:** At the same time, M&E must also be decentralized. Each line ministry and government department should have its own dedicated M&E unit responsible for monitoring its specific programs and projects. These units report to the central body, ensuring a coordinated and standardized approach across government. This structure ensures that M&E is not just a top-down mandate but is also integrated into the operational reality of service delivery.
- **Defined Roles and Responsibilities:** A key function of this structure is to clearly define the roles and responsibilities of all actors—from the central M&E body that sets policy to the line ministry units that collect data, and the civil society organizations that provide independent oversight.

3.3.4 Adequate Human and Financial Capacity

No system, however well-designed, can function without the right people and resources. Institutionalization requires a long-term commitment to building both human and financial capacity.

- **A Professional M&E Cadre:** This involves developing a permanent, professional cadre of M&E specialists within the civil service. This means offering training, creating clear career paths for M&E practitioners, and providing ongoing professional development. The aim is to move from a reliance on external consultants to an in-house expertise that is steeped in the country's public sector realities.

- **Investing in Technology and Systems:** Institutionalization also means investing in the technical infrastructure required for a modern M&E system. This includes national data systems, online M&E portals, and performance dashboards that allow for the real-time collection, analysis, and visualization of data, making M&E information more accessible and useful.

3.3.5 Fostering a Culture of M&E Use and Learning

Ultimately, the most important characteristic is not the existence of a system but its use. A truly institutionalized system is one where there is a culture of demanding, supplying, and using M&E data for decision-making and learning.

- **Leadership Buy-in:** This culture is championed from the top, with political and administrative leaders actively demanding and using M&E information to guide their policy and management decisions.
- **Transparency and Public Engagement:** An institutionalized system is transparent. It makes M&E reports and data publicly available through open data portals, which strengthens public accountability and empowers citizens to participate in governance. The open sharing of M&E findings helps build trust and makes the government more responsive to the needs of its people.
- **Learning and Adaptability:** A mature M&E system supports organizational learning. Findings from evaluations are not used to blame, but to understand what worked and what didn't, leading to policy adjustments and program improvements. This creates a feedback loop that allows the

government to become a smarter, more agile, and more effective institution.

Institutionalization of M&E is a comprehensive and systemic endeavour that goes far beyond the simple act of monitoring and evaluating a project. It is a commitment to transforming the very way a government operates, making it more results-oriented, accountable, and transparent for the long term.

3.4 Theories and Frameworks of Institutional Change

The institutionalization of Monitoring and Evaluation in Africa's public sector is, at its core, an exercise in institutional change. It is not simply a technical project but a complex process of transforming the rules, norms, and practices that govern public administration. To understand the profound challenges and opportunities of this transformation, we must turn to established theories of institutional change. These frameworks provide a lens through which to analyse why reforms succeed or fail, why some governments embrace M&E while others resist it, and how a seemingly rigid system can be transformed.

3.4.1 Neo-Institutional Theory and Its Relevance

One of the most powerful frameworks for understanding public sector reform is Neo-Institutional Theory. This theory posits that organizations, including government ministries, operate not just based on rational efficiency but also based on the social, political, and cultural norms of their environment. It helps us understand why governments often look and act alike, a phenomenon known as **institutional isomorphism**. Neo-institutionalism identifies three main mechanisms through which this change occurs:

- **Coercive Isomorphism:** This is change driven by formal or informal pressure from other, more powerful organizations. In the context of Africa's public sector, coercive isomorphism has been a major driver of M&E adoption. International financial institutions (IFIs) like the World Bank and the International Monetary Fund (IMF), as well as bilateral donors, have for decades required governments to establish M&E frameworks as a condition for receiving loans and development assistance. This pressure forced many governments to "do" M&E, even if they lacked internal demand or political will. While it may not lead to deep-seated institutionalization, this external pressure has been instrumental in placing M&E on the policy agenda. The adoption of specific M&E frameworks, such as the Results-Based Management (RBM) model, by numerous African countries can be directly traced to this coercive influence.

- **Mimetic Isomorphism:** When faced with uncertainty or a lack of clear solutions, organizations often copy the models of others that are perceived as successful or legitimate. This is mimetic isomorphism. In Africa, this has been a powerful force. A government looking to reform its public sector might look at a peer nation, such as South Africa or Uganda, that has a well-regarded national M&E system and decide to emulate it. This reduces the risk and uncertainty associated with designing a new system from scratch. However, a major pitfall of this approach is the "one-size-fits-all" trap: a system that works in one country may not be suitable for another due to different political cultures, institutional histories, and resource constraints.
- **Normative Isomorphism:** This type of change is driven by the professionalization of a field. It occurs when a new group of professionals (in this case, M&E specialists) develops and promotes a common set of norms, values, and practices. This has been a growing force on the continent with the rise of national and pan-African M&E professional associations,

such as the **African Evaluation Association (AfrEA)** and various national voluntary organizations for professional evaluation (VOPEs). These professional bodies advocate for M&E standards, provide training, and foster a shared sense of professional identity and purpose, thereby creating a "bottom-up" demand for a more institutionalized approach to M&E.

3.4.2 The Political Economy of Institutional Change

Institutional change, particularly in the public sector, is not a politically neutral process. It is fundamentally a struggle over power, resources, and influence. The political economy framework is crucial for understanding that the institutionalization of M&E can be a deeply contested process.

- **The Role of Elites and Vested Interests:** M&E, by its very nature, is a tool for transparency and accountability. It introduces a system of performance measurement that can threaten the power and interests of established elites and bureaucrats who benefit from the status quo. These actors may resist M&E institutionalization because it exposes inefficiency, makes corruption more difficult to hide, and requires them to be held accountable for results rather than just for following procedures. The lack of genuine political will to implement M&E, despite public pronouncements, is often a sign of this internal resistance.
- **Political Will as a Catalyst:** While resistance is common, genuine political will from the highest levels of government is a powerful catalyst for institutional change. A president, prime minister, or finance minister who is a true champion of M&E can use their authority to overcome bureaucratic inertia, allocate necessary resources, and hold the civil service accountable for using M&E findings. The location of a central

M&E unit in the office of the president or prime minister, as seen in Uganda and South Africa, is a strategic choice designed to signal this political commitment and give the system the necessary authority.

- **The Distribution of Costs and Benefits:** The political economy framework also highlights that institutional change has distributional consequences. An M&E system may benefit citizens by improving public services, but it may impose costs on public officials who are now under greater scrutiny. Understanding who benefits and who loses from M&E reform is essential for designing a strategy that can build a coalition of support while mitigating the inevitable resistance.

3.4.3 The Role of Change Agents and Leadership

While broader theories help explain the context of change, institutional change is often driven by human agency. The concept of a change agent or institutional entrepreneur is central here. These are individuals or small groups who have the vision, skill, and political savvy to push for new institutional arrangements.

- **Visionary Leadership:** In the M&E context, a change agent could be a reform-minded senior civil servant, a powerful minister, or a respected academic who tirelessly advocates for the value of M&E. They can articulate a compelling vision for a more results-oriented government and build coalitions of support across different parts of the public sector and civil society.
- **Building Consensus and Coalitions:** Change agents don't work in isolation. They are adept at building coalitions of support by linking the M&E agenda to the self-interests of different stakeholders. For example, they might convince the Ministry of Finance that M&E will help them allocate

resources more effectively, or they might engage civil society by promising greater transparency. This ability to build a broad base of support is crucial for overcoming resistance and sustaining the momentum for reform over time.

3.4.4 Punctuated Equilibrium Theory

The theories discussed so far often assume that institutional change is a gradual, incremental process. However, the Punctuated Equilibrium Theory offers a powerful alternative explanation, suggesting that institutions are often characterized by long periods of stability (equilibrium) that are interrupted by short, dramatic bursts of rapid change (punctuations).

- **Equilibrium and Stability:** In the context of public sector M&E, equilibrium can be the "business as usual" of a bureaucratic system where change is slow, incremental, and difficult to achieve. Institutional friction—the inertia of rules, norms, and existing power structures—resists new ideas like M&E.
- **Punctuation and Windows of Opportunity:** A "punctuation" is a crisis or a focusing event that creates a sudden window of opportunity for radical change. In Africa, such events might include:
    - A severe economic crisis that exposes the inefficiency and waste in public spending.
    - A major corruption scandal that erodes public trust and creates a public outcry for accountability.
    - A change in government that brings in a new political leadership with a mandate for reform.
- **Leveraging the Moment:** These crises can dismantle the old equilibrium and create the political space for change agents to push through previously stalled reforms. The sudden adoption

of a national M&E system in a country following a governance crisis is a classic example of this theory in action. A long period of slow progress can be dramatically accelerated by a single, seismic event.

3.4.5 Synthesizing the Frameworks

No single theory can fully explain the complex journey of M&E institutionalization in Africa. Instead, a comprehensive understanding requires a synthesis of these frameworks. The coercive pressures of international donors may place M&E on the agenda (Neo-Institutionalism), but a lack of political will from elites (Political Economy) can prevent its full adoption. A change agent might patiently build a coalition for M&E (Leadership), but it may take a major crisis to create the necessary political space for institutionalization to occur (Punctuated Equilibrium).

By understanding these theoretical underpinnings, we can move beyond a simplistic view of M&E as a technical fix. We can see it for what it truly is: a powerful and often contested tool for institutional change that has the potential to reshape the landscape of governance in Africa.

Chapter 4

The Legal and Policy Framework for M&E

Introduction

The institutionalization of Monitoring and Evaluation (M&E) is a journey from an ad-hoc, discretionary activity to a core, non-negotiable function of government. This transformation cannot be sustained by good intentions alone; it must be anchored in a robust and binding legal and policy framework. Without a clear mandate and authority, M&E systems are vulnerable to political shifts, budgetary cuts, and bureaucratic resistance. This chapter explores the foundational role of national constitutions, laws, and policies in mandating M&E, providing the system with the legitimacy, durability, and teeth it needs to be effective.

4.1 The Authority of Mandate: From Choice to Obligation

The most significant step in institutionalizing M&E is a government's decision to move from "doing" M&E as a choice—often under external pressure—to "doing" M&E as a legal obligation. This shift is what provides M&E systems with their power. A law or policy formalizes the roles and responsibilities, establishes a common language, and creates a system that is resilient to changes in political leadership. This framework acts as a public commitment to accountability, signalling to citizens and stakeholders that the government is serious about results.

4.1.1 The Role of National Constitutions: The Foundational Pillar

While national constitutions rarely contain explicit mention of "monitoring and evaluation," they often provide the ultimate legal authority for M&E through their foundational principles. Many modern African constitutions, born from struggles for independence and democratic governance, contain powerful clauses that implicitly mandate the use of M&E.

These foundational principles include:

- **Accountability and Transparency:** Constitutions often state that public power is held in trust for the people and that the government must be accountable and transparent in its actions. M&E, by its nature, is the primary technical mechanism for achieving this.
- **Efficient and Effective Use of Public Resources:** Many constitutions include clauses that require the government to use public funds efficiently and for the public good. An M&E system is essential for proving that resources are being used effectively and not wasted on non-performing programs.
- **Good Governance:** The broad principle of good governance, which is a pillar of many constitutions, encompasses elements of performance, results, and citizen engagement—all of which are supported by a strong M&E system.

For example, a country's constitution might declare that all public officials must be held accountable for their performance. An M&E law or policy can then be enacted to define what performance means, how it will be measured, and what the consequences of poor performance will be. The constitution, therefore, provides the moral and legal high ground from which a government can mandate M&E.

4.1.2 The Power of M&E Laws and Acts: The Direct Mandate

The most direct and powerful way to institutionalize M&E is through a dedicated law or an act of parliament. A law carries the full force of the state and is more difficult to reverse than a simple administrative directive. Such legislation typically outlines the following:

- **Creation of a National M&E Body:** The law can officially establish a central M&E unit, such as a department or directorate, and specify its mandate, powers, and reporting lines. This formalizes its existence and protects it from being disbanded by a new administration.

- **Mandatory M&E for all Public Entities:** An M&E act can require all ministries, departments, and public entities to establish their own M&E units, develop performance indicators, and report on their results. This ensures a "whole-of-government" approach rather than a siloed, project-by-project system.
- **Reporting Requirements:** The law can specify a clear framework for how M&E findings will be reported. This includes the content, frequency, and recipients of reports, such as a mandatory annual M&E report to be submitted to Parliament and made public.
- **Integration with Financial Management:** A key element of a strong legal framework is the link to public finance. In Ghana, the Public Financial Management Act (PFMA) is a prime example of this. The Act mandates that all public entities must prepare strategic plans with clear objectives and indicators and submit annual reports on their performance against those targets. This powerful legal mandate makes M&E a non-negotiable part of a government's financial responsibilities.

A legal mandate ensures that M&E is not just a passing trend but a permanent function of the state.

4.1.3 National M&E Policies and Strategies: The Operational Roadmap

While laws provide the "what," national M&E policies and strategies provide the "how." A policy is a government's official statement of its vision, principles, and approach to M&E. It serves as an operational roadmap for how the M&E law will be implemented.

A comprehensive national M&E policy typically includes:

- **Vision and Principles:** It outlines the government's long-term vision for M&E and the core principles that will guide its practice, such as transparency, objectivity, and a focus on learning.

- **Roles and Responsibilities:** The policy clarifies the roles of all key actors in the M&E system, from the central coordinating body to line ministries, the legislature, and civil society. This prevents confusion and ensures accountability. Uganda's National Policy on Monitoring and Evaluation is a strong example of a document that provides a guiding framework for all M&E-related activities across its government.
- **Standards and Methodologies:** The policy sets out the common standards for the M&E system. This includes a standardized logical framework, a common set of indicators for national priorities, and guidelines for conducting evaluations.
- **Capacity Building:** A policy often includes a plan for building the human and technical capacity required to run the M&E system. This involves strategies for training civil servants, investing in data systems, and promoting a culture of M&E use.
- **Frameworks for Specific Sectors:** Some countries, like Kenya, have developed a National Integrated Monitoring and Evaluation System (NIMES), which is not just a policy but a comprehensive framework that includes specific guidelines for M&E in different sectors.

4.1.4 The Interplay: Building a Resilient M&E System

The most resilient and effective M&E systems are those that are supported by all three levels: a constitutional foundation, a strong legal mandate, and a clear policy roadmap.

- A **constitutional** mandate provides the ultimate authority and legitimacy.
- A **law** provides the binding, legal force that makes M&E a mandatory function.
- A **policy** provides the operational detail and strategic guidance that makes the system work in practice.

The combination of these elements creates a system that is protected from administrative whims and changes in political leadership. It ensures that M&E is not just an activity but a permanent institution that is embedded in the very fabric of government, thereby creating a solid foundation for a more accountable, transparent, and results-oriented public sector in Africa.

4.2.0 Case Studies: M&E Policies and Strategies in Select African Countries

The journey toward institutionalizing Monitoring and Evaluation (M&E) is unique to each country, shaped by its political history, governance structure, and administrative traditions. While no single model is universally applicable, a comparative analysis of countries that have made significant progress offers invaluable insights. By examining the approaches of South Africa, Uganda, and Ghana, we can uncover best practices, common pitfalls, and the strategic choices that underpin successful M&E systems in the African context.

4.2.1 Case Study 1: South Africa – A High-Level, Centralized Approach

South Africa's M&E system is arguably one of the most advanced and well-resourced in Africa, and its unique institutional design is a key reason for its success. Following the 2009 general election, the government established a centralized M&E function to drive a more performance-oriented state.

- **Key Institutional Design:** The M&E function is housed

within the Department of Planning, Monitoring, and Evaluation (DPME), which is located in the Presidency. This high-level location is a powerful statement of political will, providing the DPME with the authority to hold other government departments accountable. Its mandate is not to monitor or evaluate every project, but to oversee a National Evaluation System (NES) that coordinates, supports, and guides M&E activities across government.

- **Core Functions and Successes:** The DPME has been instrumental in building a culture of evaluation. Its core functions include:
    - **Development of a National Evaluation Plan:** The DPME regularly publishes a plan that identifies key government programs to be evaluated, ensuring that limited evaluation resources are focused on areas of national importance.
    - **Creation of an M&E Framework:** It developed a clear M&E framework that provides a common language and standards for all government departments.
    - **Oversight and Accountability:** The DPME produces and submits comprehensive performance reports on key national priorities to Cabinet and Parliament. This creates a strong mechanism for upward accountability, as ministers and departments are required to answer for their performance.
- **Challenges and Lessons:** Despite its successes, the South African model faces challenges. The high-level nature of the DPME can sometimes create a distance between the central M&E body and the operational realities of line ministries. A persistent challenge has been the full integration of M&E findings into the budget allocation process, although progress is being made. The key lesson from South Africa is that placing the M&E function at the highest level of government can

be a powerful catalyst for change, but it requires a sustained effort to embed the culture of performance at every level of the bureaucracy.

4.2.2 Case Study 2: Uganda – A Policy-Driven, Decentralized System

Uganda's approach to institutionalizing M&E is a classic example of a system built on a comprehensive and clear policy framework. The government's efforts began in the early 2000s, driven by a desire for improved public service delivery and greater accountability.

- **Key Institutional Design:** Uganda's M&E system is guided by a National Policy on Monitoring and Evaluation, which provides a clear and detailed roadmap. The policy outlines the roles and responsibilities of key institutions, including the Office of the Prime Minister (OPM), which is the central coordinating body, and the M&E units within ministries and local governments. This design promotes a decentralized system where M&E is not just a top-down mandate but is also a function of local governance.
- **Core Functions and Successes:** The Ugandan model is lauded for several key components:
    - **Emphasis on Policy:** The National M&E Policy serves as a single, comprehensive reference document that provides clarity and a common vision for all stakeholders.
    - **Stakeholder Participation:** The policy places a strong emphasis on participatory M&E, engaging civil society and the public in the oversight process.
    - **Integration with the National Development Plan:** The M&E system is directly linked to the National Development Plan (NDP), ensuring that M&E efforts are focused on tracking progress toward national priorities.

- **Challenges and Lessons:** The Ugandan model, while theoretically sound, faces challenges in practice. A key issue is the capacity gap at the local government level, where M&E units often lack the necessary technical skills and resources to effectively monitor programs. The sheer number of districts also makes a national, standardized approach difficult to manage. The key lesson from Uganda is that a clear and comprehensive policy is a necessary first step, but it must be followed by a sustained commitment to capacity building and a robust system for ensuring compliance at all levels of government.

4.2.3 Case Study 3: Ghana – A Legislative-Backed Framework

Ghana's approach to M&E is distinctive because it is firmly anchored in a broader legal framework that makes M&E a non-negotiable part of a government's financial responsibilities. This provides M&E with a durability that is less susceptible to political whims.

- **Key Institutional Design:** The legal foundation for M&E in Ghana is the Public Financial Management Act (PFMA) of 2016. This legislation mandates that all public entities to prepare strategic plans with clear objectives and performance indicators. Furthermore, it requires them to submit annual reports to Parliament detailing their performance against these targets. This powerful legislative backing elevates M&E from a mere policy to a legal duty.
- **Core Functions and Successes:** The Ghanaian model's key strengths lie in its:
    - **Legal Authority:** The PFMA provides a strong and binding legal mandate for M&E, ensuring that ministries and agencies must comply.

   - **Integration with Financial Management:** By linking M&E to the financial management process, the government ensures that M&E is not a separate function but an integral part of how public funds are managed and accounted for. This creates a powerful incentive for compliance.
   - **Oversight by Parliament:** The requirement to submit performance reports to Parliament creates a strong mechanism for legislative oversight, thereby strengthening upward accountability.
- **Challenges and Lessons:** While the PFMA provides a strong legal mandate, a challenge has been the need for a more comprehensive operational roadmap. While the law mandates M&E, the need for a national M&E policy is becoming more apparent to provide a common framework, standardize methodologies, and build the necessary capacity across all public entities. The key lesson from Ghana is that anchoring M&E in a legal framework, particularly one related to financial management, provides a durable foundation for accountability, but it must be complemented by a clear operational policy to be fully effective.

4.2.4 Comparative Analysis and Lessons Learned

By comparing these three countries, we can draw several key conclusions:

- **Political Will Matters:** All three systems, whether centralized like South Africa's or decentralized like Uganda's, were initiated with strong political will. This buy-in from the highest level of government is a prerequisite for any successful M&E institutionalization effort.
- **The Power of a Clear Framework:** Each country's model is defined by a clear framework—a department in the

Presidency, a national policy, or a legal act. A clear framework is essential for providing direction, assigning responsibilities, and ensuring consistency.

- **Challenges of Implementation:** All three countries face common challenges, including gaps in technical capacity, particularly at the sub-national level, and the ongoing struggle to fully integrate M&E evidence into the budgeting process.
- **No Single "Right" Model:** There is no single correct path to M&E institutionalization. South Africa's centralized, top-down approach is effective for driving a national agenda, while Uganda's decentralized, policy-driven model is effective for promoting participation. Ghana's legal approach provides a strong foundation for accountability. The choice of model depends on a country's unique governance structure, political culture, and reform goals.

The experiences of these countries serve as a vital guide for other African nations seeking to build their own M&E systems. They demonstrate that institutionalization is a long-term, complex process that requires a strategic blend of political will, legal frameworks, and a sustained commitment to building capacity and fostering a culture of performance.

4.3.1 Ensuring M&E is Not Optional, But a Required Function of Government

For Monitoring and Evaluation (M&E) to be truly institutionalized, it must move beyond being a project management tool or an optional administrative exercise. It must become a non-negotiable, legally required function of government. A supportive legal and policy environment is the bedrock of this transformation, providing M&E with the legitimacy, authority, and durability it needs to withstand political and bureaucratic pressures. This section discusses the imperative for a legal mandate and outlines the key components and processes required to establish such a framework in the African public sector.

4.3.2 The Imperative for a Legal Mandate

In the absence of a legal mandate, M&E is often vulnerable. A new political administration can easily dismantle an M&E system that was established by a predecessor's administrative directive. A legal mandate, however, provides M&E with a resilience and permanence that is far more difficult to reverse. It transforms M&E from a "nice-to-have" into an essential public function, just like budgeting or auditing.

A legal mandate is critical for several reasons:

- **Provides Legitimacy:** A law or an act of parliament gives M&E a formal standing and authority, signalling to all government entities that it is a serious and binding function.
- **Ensures Durability:** By being enshrined in law, M&E systems are protected from the whims of changing political leadership.
- **Strengthens Accountability:** A legal mandate provides the basis for holding public officials and institutions accountable. When a law requires a ministry to report on its performance, it creates a powerful legal mechanism for oversight bodies, such as Parliament, to demand answers.

4.3.3 Key Components of a Supportive Legal Framework

A truly supportive legal environment for M&E is a layered system, with a mandate at multiple levels of government and within various pieces of legislation.

1. The Constitutional Foundation: The highest level of legal authority for M&E often resides in the national constitution. While constitutions do not explicitly mention M&E, they provide the moral and legal foundation by championing principles such as:

- **Public Accountability:** The constitution establishes that public power is a trust from the people, requiring a mechanism to hold government accountable.

- **Efficient Use of Public Resources:** Many constitutions mandate that public funds must be used effectively and for the public good, which requires an M&E system to verify.

These constitutional principles provide the ultimate legal justification for any subsequent M&E law or policy.

2. A Dedicated M&E Law or Act: A dedicated law is the most direct way to formalize M&E. Such a law would typically:

- **Establish a Central M&E Body:** Legally create a national M&E department or agency, defining its mandate, reporting lines, and relationship with other parts of government.
- **Define Roles and Responsibilities:** Mandate that all ministries, departments, and agencies (MDAs) must have M&E functions and staff.
- **Require Mandatory Reporting:** Prescribe a clear framework for reporting M&E findings to a central body, the legislature, and the public.

3. Integration into Broader Legislation: A highly effective strategy is to embed M&E requirements into other key pieces of legislation, which provides a powerful and pragmatic anchor.

- **Public Financial Management (PFM) Acts:** This is one of the strongest anchors. As seen in Ghana, the PFMA of 2016 requires all public entities to prepare strategic plans with clear performance indicators and to report on their performance against these targets. This legally ties M&E directly to a government's financial responsibilities.
- **Public Service Acts:** M&E can be linked to the performance

management of individual civil servants, making it a condition of their employment and promotion.

- **Local Government Acts:** At the sub-national level, laws such as the Municipal Systems Act in South Africa can require municipalities to implement performance management systems with key performance indicators to measure their developmental interventions.

4.3.4 The Process of Establishing the Legal Framework

Creating a legal framework is a structured process that moves from a concept to a binding legal reality.

- **Step 1: Policy Development and Stakeholder Consultation:** The process often begins with the development of a national M&E policy, which serves as a blueprint. This policy should be a product of a broad consultative process, including input from civil society, academia, the private sector, and parliamentarians. This ensures buy-in and makes the eventual law more legitimate and robust.
- **Step 2: Drafting the Bill:** Based on the national policy, a bill is drafted by legal experts, detailing the specific mandates, structures, and responsibilities.
- **Step 3: Legislative Passage:** The bill is then submitted to Parliament, where it undergoes a rigorous legislative process of debate and review before being passed into law.
- **Step 4: Budgeting and Implementation:** The legal framework must be supported by a dedicated budget to fund the new M&E structures and functions. Without adequate resources, a legal mandate is meaningless.

4.3.5 From Legal Mandate to Culture of Accountability

A legal framework is a powerful enabler, but it is not a magic bullet. The ultimate goal is for the legal mandate to foster a culture of compliance and accountability. This requires:

- **Enforcement:** The law must be enforced. Parliamentary oversight committees and audit institutions must use the legal mandate to demand high-quality M&E reports and to hold the executive accountable for the results.
- **Capacity Building:** The legal framework must be supported by a sustained effort to build the technical capacity of civil servants to conduct M&E effectively.
- **Political Will:** The legal environment is most effective when it is continuously supported by strong political will from the highest levels of government.

In summary, creating a supportive legal environment is the most critical step in moving M&E from a temporary project to a permanent, core function of government. It provides the system with the authority to demand results, the legitimacy to hold officials accountable, and the resilience to survive over the long term, thereby providing a durable foundation for good governance in the African public sector.

Chapter 5

Structural and Organizational Arrangements for M&E

Introduction

The journey toward an institutionalized M&E system is not merely about having the right laws or a clear policy; it is about making strategic organizational choices that give the system the power and influence to effect real change. At the heart of this decision-making process lies the question: "Where should the central M&E function be located within the machinery of government?" The answer to this question is not an administrative detail but a profound strategic choice that shapes the M&E system's authority, independence, and relationship with the rest of the public sector. This chapter provides a comprehensive discussion on the different models for locating the central M&E body, analysing the strengths and weaknesses of each, and drawing lessons from across the African continent.

5.1 The Strategic Importance of Location

The location of a country's central M&E body is a direct reflection of its government's priorities. Is the primary goal of M&E to enhance political accountability? To improve the efficiency of resource allocation? Or to inform long-term national development plans? The answer to these questions dictates where the M&E unit is most likely to be effective. The location determines its access to power, its ability to influence decision-making, and its capacity to enforce its mandate across a diverse and often resistant bureaucracy.

While there is no single "right" place to locate a central M&E body, three main models have emerged in African public sector governance: in the Office of the President or Prime Minister, in the Ministry of Finance, or in a national Planning Commission. Each model comes with a distinct set of advantages and disadvantages.

5.2 M&E in the Office of the President or Prime Minister

This model places the M&E function at the highest echelon of government, often as a department or a directorate within the Office of the President or Prime Minister. This is a deliberate strategic choice aimed at leveraging political authority to drive a "whole-of-government" results agenda.

Case Study: South Africa's Department of Planning, Monitoring, and Evaluation (DPME)

South Africa is a prime example of this model. The DPME was established in the Presidency in 2009 with a clear mandate to improve the government's performance and accountability.

Advantages:

- **Unparalleled Political Authority:** Located in the Presidency, the central M&E body has the political weight to demand compliance from other ministries and agencies. A minister or director-general from the DPME can speak with the authority of the President, making it difficult for line ministries to ignore M&E findings. This authority is crucial for overcoming bureaucratic inertia.
- **A "Whole-of-Government" Perspective:** From this vantage point, the M&E body can coordinate monitoring and evaluation across the entire government. It is uniquely positioned to identify cross-cutting issues, track progress on national priorities, and ensure that all ministries are aligned with the government's overarching strategic plan.
- **Direct Link to Executive Decision-Making:** M&E findings can be presented directly to the President and Cabinet, ensuring that evidence is a key input into the highest-level policy and resource allocation decisions. This model makes M&E a central tool for executive accountability.

Disadvantages:

- **Risk of Politicization:** The close proximity to political power can also be a weakness. M&E findings might be susceptible to being filtered, spun, or ignored for political reasons. There is a risk that M&E becomes a tool to showcase political successes rather than an independent and objective mechanism for learning and accountability.

- **Potential Lack of Technical Independence:** While the central M&E body may have a strong mandate, its technical independence can be challenged. The perception that M&E is a political tool can lead to a lack of trust from the public and civil society.
- **Distance from Budgetary Processes:** A major drawback is the potential distance from the Ministry of Finance and the day-to-day realities of budget allocation. While a high-level mandate can be powerful, if the M&E findings are not directly linked to budget decisions, their influence can be limited.

5.3 M&E in the Ministry of Finance or Treasury: The Performance Budgeting Model

This model anchors the central M&E function within the Ministry of Finance, the government department with ultimate control over the budget. This is a strategic choice that views M&E as an instrument for public financial management and performance-informed budgeting.

Case Study: Zambia's M&E Efforts in the Ministry of Finance

In some countries, the Ministry of Finance plays a central role in driving M&E, often through its control of the national budget. It is a powerful lever for change, as it can link M&E findings directly to resource allocation decisions.

Advantages:

- **Direct Link to the Budget Process:** This is the most significant advantage. By placing M&E in the Ministry of Finance, the findings are directly integrated into the annual budget cycle. M&E data can be used to justify budget increases for well-performing programs and to cut or reallocate funds from underperforming ones. This creates a powerful incentive for line ministries to take M&E seriously.

- **Emphasis on Efficiency and Value for Money:** M&E in this context often has a strong focus on financial accountability and cost-effectiveness. The M&E unit is uniquely positioned to ensure that public funds are not only being spent as planned but are also achieving the desired results.
- **Strong Technical Capacity:** Ministries of Finance are often among the most technically capable government departments, with a strong focus on data analysis, economics, and quantitative methods. This technical expertise can provide a strong foundation for a robust M&E system.

Disadvantages:

- **Focus on Fiscal Compliance:** A potential drawback is that M&E can become overly focused on fiscal compliance rather than on a broader assessment of program outcomes and impact. The emphasis may be on "did you spend the money?" rather than "did the program achieve its goals?"
- **Lack of Independence:** M&E in this context can be seen as a tool for the Ministry of Finance to exert control over other ministries, potentially leading to resistance and a lack of cooperation. The independence of M&E findings can be compromised if they are used to justify pre-determined budget cuts.
- **Limited Scope:** This model may struggle to provide a comprehensive "whole-of-government" picture, as its primary focus is on financial performance rather than on broader

strategic oversight.

5.4 M&E in a Planning Commission or Ministry: The Development-Focused Model

This model places the M&E function within the national Planning Commission or a Ministry of Planning. This is a strategic choice that views M&E as an integral part of the planning cycle, a tool for ensuring that national development goals are being met.

Case Study: Uganda's M&E Framework

While Uganda's M&E system is coordinated by the Office of the Prime Minister, it is heavily linked to the national planning framework, with the Ministry of Finance, Planning, and Economic Development playing a key role. This illustrates the close relationship between M&E and planning.

Advantages:

- **Strong Link to National Development Plans:** Locating M&E in a Planning Ministry ensures a direct and powerful link between M&E findings and the country's long-term vision. M&E data can be used to inform the next planning cycle, ensuring that future interventions are based on a strong evidence base.
- **Focus on Strategic Outcomes:** This model encourages M&E to look beyond outputs and focus on the short, medium-, and long-term outcomes and impact of programs. The M&E body can ensure that all government activities are contributing to the country's strategic vision.
- **Emphasis on Learning and Improvement:** A planning-focused M&E system often has a strong emphasis on using M&E findings for learning. It is less about accountability and more about using evidence to build better programs and policies in the future.

Disadvantages:

- **Potential Lack of Authority:** In many African countries, a Planning Ministry may be a weaker ministry without the political authority of the Presidency or the financial clout of the Ministry of Finance. This can make it difficult for the M&E body to enforce its mandate across other ministries.
- **Distance from Budgetary Processes:** As with the Presidency model, M&E in a Planning Ministry can be disconnected from the budget process. If the M&E findings are not directly linked to resource allocation, their influence may be limited to the planning cycle.
- **Risk of Ineffectiveness:** If the M&E function is seen as an academic or technical exercise without real-world consequences, it can lead to a lack of buy-in from other government departments.

Comparative Analysis and the "Best Fit" Model

M&E Location	Key Advantages	Key Disadvantages
Presidency	High political authority, whole-of-government view, direct link to executive decisions.	Risk of politicization, distance from budget, potential lack of technical independence.
Ministry of Finance	Direct link to the budget, focus on efficiency and value for money, strong technical capacity.	Overly focused on fiscal compliance, lack of independence from finance, narrower scope.
Planning Commission	Strong link to national development plans, focus on strategic outcomes, emphasis on learning.	Potential lack of authority, distance from the budget, risk of being seen as an academic exercise.

In summary, there is no one-size-fits-all model for locating a central M&E body. The "best" location is a matter of "best fit," a strategic decision that depends on a country's unique governance structure, political culture, and priorities. A government that prioritizes political accountability and national oversight may choose the Presidency model. A government focused on fiscal discipline and performance budgeting may choose the Ministry of Finance. And a government focused on long-term national development may choose a Planning Commission.

5.2 The Importance of a Central Coordinating Body and Its Relationship with Line Ministries

The institutionalization of Monitoring and Evaluation (M&E) is a whole-of-government endeavour that requires a delicate balance between centralized coordination and decentralized implementation. While line ministries and agencies are the primary implementers of M&E—collecting data, monitoring projects, and conducting sectoral evaluations—a central coordinating body is the "brain and nervous system" of the entire system. Its role is to provide strategic direction, ensure consistency, and synthesize disparate data into a coherent national picture. The nature of its relationship with line ministries is the single most critical determinant of the M&E system's effectiveness and long-term sustainability.

5.2.1 The Core Functions of a Central M&E Coordinating Body

A central M&E body, regardless of its location (as discussed in the previous section), serves as the architect and steward of the national M&E system. Its core functions can be broadly categorized as:

- **Setting Standards and Methodologies:** To ensure that M&E data is comparable and credible across government, the central body develops and institutionalizes common standards and methodologies. This includes a standardized logical framework, a common set of indicators for national priorities, and clear guidelines for conducting evaluations. Without this, each ministry would operate in a silo, producing data that cannot be aggregated for a national picture.
- **Providing Technical Support and Capacity Building:** Line

ministries often have varying levels of M&E capacity. The central body's role is to act as a service provider, offering training, technical backstopping, and access to advanced M&E tools. This support is crucial for bridging capacity gaps and ensuring that all parts of the government can effectively implement the national framework.

- **Synthesizing a National Picture:** The central body is uniquely positioned to aggregate data from all ministries and agencies to produce a comprehensive, "whole-of-government" report on progress toward national development goals. It identifies cross-cutting issues, helps resolve implementation problems that are beyond the control of a single ministry, and provides a national narrative of success and challenges.
- **Oversight and Quality Assurance:** The central body is responsible for ensuring the quality and integrity of M&E data. It can do this through data verification visits, conducting quality audits of departmental M&E reports, and commissioning independent evaluations of key government programs to ensure objectivity and credibility.

5.2.2 The Dynamic Relationship: A Blend of Service and Oversight

The relationship between the central M&E body and the line ministries is a dynamic one, a constant balancing act between being a collaborative partner and a strict overseer.

The Central Body as a Service Provider:

In this role, the central body's primary objective is to empower line ministries. It does not dictate but rather facilitates. This involves:

- **Training and Mentorship:** Organizing training workshops for ministerial M&E focal points and program managers.
- **Sharing Best Practices:** Disseminating lessons learned from

successful evaluations and sharing effective M&E tools and techniques.

- **Providing a Help Desk Function:** Being available to answer technical questions and provide guidance on a day-to-day basis.

This collaborative approach is essential for building ownership and ensuring that ministries do not see M&E as an external imposition. The goal is to build a relationship of trust where ministries are comfortable sharing their challenges and are confident that the central body is there to help them succeed.

The Central Body as an Oversight Body:

In its oversight role, the central body uses its authority to ensure compliance with the national M&E framework. This is the accountability dimension of the relationship. This involves:

- **Enforcing Reporting Requirements:** Ensuring that ministries submit their M&E reports on time and in the required format.
- **Performance Reviews:** Conducting regular performance reviews with ministers and their senior management teams to discuss progress, challenges, and remedial actions.
- **Holding Ministries Accountable:** Using M&E data to hold ministries accountable for their performance, and escalating issues of non-compliance to the highest levels of government, such as Cabinet or Parliament.

The effectiveness of this oversight function is directly linked to the central body's location. A body in the Presidency, like South Africa's Department of Planning, Monitoring, and Evaluation (DPME), has significant political clout to enforce its mandate. A body in a weaker ministry, however, might struggle to hold more powerful ministries accountable.

5.2.3 Case Study: South Africa's DPME and Its Relationship with Line Ministries

South Africa's DPME provides a powerful illustration of a central-ministry relationship with a strong oversight mandate. Its vision is to "instil in all public service departments a performance-driven management culture."

The DPME's relationship with line ministries is built on a clear framework of accountability:

- **Performance Agreements:** Ministers sign Performance Agreements with the President, which are linked to national priorities. The DPME is responsible for tracking progress against these agreements.
- **Reporting and Analysis:** Line ministries are required to submit regular performance reports to the DPME. The DPME's role is not just to receive these reports but to analyse their content, challenge the data, and provide an independent assessment of ministerial performance.
- **Verification and Interventions:** The DPME conducts verification visits to project sites and engages directly with departments on their performance. When challenges arise, the DPME has the authority to intervene on behalf of the President and Cabinet to help resolve them.

This model, while effective, is a constant struggle against a "silo mentality" and bureaucratic resistance. It is a delicate balance of providing support while maintaining the authority to hold ministries accountable for their performance.

5.2.4 Common Challenges and Best Practices

The relationship between the central body and line ministries is fraught with potential challenges that can derail even the best-designed M&E system.

Common Challenges:

- **Resistance from Line Ministries:** Ministries may resist the central M&E body if they perceive it as an external auditor rather than a supportive partner. This can lead to a lack of cooperation and the submission of inaccurate or incomplete data.
- **Capacity Gaps:** A central body may have a clear mandate, but if line ministries lack the technical capacity to collect quality data and conduct evaluations, the system will fail.
- **"Tick-Box" Mentality:** Ministries may see M&E as a compliance exercise rather than a tool for learning and improving. They may simply "tick the box" by submitting reports without using the data for their own decision-making.

Best Practices for Success:

- **Build a Collaborative Culture:** The central body should actively foster a collaborative culture by emphasizing its role as a partner in improving performance. This can be done through joint workshops, peer-learning events, and a focus on sharing success stories.
- **Provide Clear Incentives:** M&E should not be a thankless task. Incentives, such as linking M&E results to performance bonuses for civil servants, can foster a culture of evidence use.
- **Start Small and Build Momentum:** The central body should focus its initial efforts on a few key, high-visibility projects to demonstrate the value of M&E before scaling it up across the entire government.

- **Embed M&E in Key Processes:** A crucial best practice is to embed M&E requirements in existing processes, such as budgeting and strategic planning. This makes it a non-negotiable part of a ministry's core function rather than an add-on.

The central M&E body is the engine of a national M&E system, but its effectiveness is entirely dependent on its relationship with line ministries. By successfully balancing its roles as a service provider and an oversight body, a central M&E body can foster a culture of accountability and evidence-based decision-making, transforming M&E from a technical exercise into a powerful tool for good governance.

5.3 Developing M&E Units within Ministries, Departments, and Agencies (MDAs)

While a central M&E coordinating body provides the strategic direction and national oversight, the operational heart of any M&E system lies within the individual Ministries, Departments, and Agencies (MDAs). It is at this level that policies are implemented, projects are executed, and services are delivered. The establishment of dedicated M&E units or functions within MDAs is, therefore, a critical step in institutionalizing M&E, transforming a national mandate into a daily operational reality. This section provides a comprehensive discussion on the purpose, key functions, common challenges, and best practices for developing these vital M&E units.

5.3.1 The Mandate and Purpose of MDA-Level M&E Units

The primary purpose of an M&E unit within an MDA is to serve as the internal engine for evidence-based decision-making. These units are responsible for ensuring that the national M&E policy is effectively implemented within their specific sector or mandate. Their work is essential for:

- **Informing Internal Management:** The M&E unit provides the ministry's leadership with timely and accurate information on program progress, resource utilization, and performance. This data allows managers to make course corrections, reallocate resources, and address implementation challenges before they escalate.
- **Facilitating Accountability:** MDA M&E units are the primary source of data for both upward and downward accountability. They are responsible for reporting their performance to the central M&E body and, ultimately, to Parliament and the public. At the same time, they enable downward accountability by providing information to citizens and stakeholders on the delivery of services.
- **Supporting Strategic Planning:** By providing evidence on "what works and what doesn't," the M&E unit plays a critical role in informing the MDA's strategic planning and annual budgeting processes. This ensures that future plans are based on past performance, not just on assumptions.

5.3.2 Key Functions of an MDA-Level M&E Unit

A functional M&E unit within an MDA performs a range of crucial tasks that make the national M&E system operational. Its functions can be grouped into the following categories:

- **Strategic and Annual Planning:** The unit collaborates with ministry leadership to develop a robust M&E plan that is integrated into the MDA's strategic and annual plans. This includes defining clear objectives, key performance indicators (KPIs), and a results framework that aligns with national priorities. As highlighted in Nigeria's National M&E Policy, the central body works with MDAs to develop "results-focused, key performance indicators and clearly defined performance targets."

- **Data Management and Quality Assurance:** This is the core technical function. The unit designs and manages the ministry's data collection tools, M&E databases, and information systems. It is responsible for ensuring the accuracy, reliability, and timeliness of the data collected, often conducting data verification exercises to maintain integrity.
- **Monitoring and Reporting:** The unit oversees the day-to-day monitoring of projects and programs, tracks progress against performance targets, and prepares regular progress reports for both internal management and the central M&E body. This includes preparing the mandatory reports that are a requirement in many M&E policies and laws.
- **Evaluation Coordination:** The M&E unit is responsible for coordinating program and project evaluations within the ministry. This includes developing terms of reference, managing the work of external consultants, and ensuring that the evaluation findings are of high quality and are disseminated to relevant stakeholders.
- **Capacity Building:** As the internal M&E champion, the unit provides training, mentorship, and support to program managers and other staff within the ministry. This function is vital for fostering an M&E culture and ensuring that all staff understand their role in the system.

5.3.3 Challenges in Establishing and Strengthening MDA-Level M&E Units

Despite their importance, the establishment and strengthening of M&E units within MDAs are often met with significant challenges. These are the same challenges that can undermine an otherwise well-designed national M&E system.

- **Inadequate Capacity and Skills:** One of the most common challenges is a lack of technical capacity. Staff assigned to M&E units often lack the necessary skills in data analysis, report writing, and evaluation methodologies. A study on South African municipalities, for example, highlighted the persistent challenge of "inadequate understanding of M&E" and a lack of specialized personnel.
- **Budgetary Constraints:** M&E units frequently lack a dedicated budget, forcing them to rely on project funds or external support. This makes it difficult to conduct regular monitoring visits, manage data systems, or commission independent evaluations. Without a dedicated budget line, the M&E function is seen as an add-on rather than a core function.
- **Resistance and a "Tick-Box" Mentality:** Line ministries may resist the M&E function if it is seen as an external burden or a tool for central government to exert control. This can lead to a "tick-box" mentality, where M&E reports are produced for the sake of compliance, but the data is not actually used for decision-making. As a study on Ghana's M&E system pointed out, a common challenge is M&E being tied to "nominal compliance" rather than "evidence-based decision making."
- **Lack of Leadership Buy-in:** The success of an MDA-level M&E unit depends heavily on the support of the ministry's leadership. If the Minister, Director-General, or senior management do not value M&E, the unit will lack the authority and resources to be effective.

5.3.4 Best Practices for Developing Effective MDA-Level M&E Units

Overcoming these challenges requires a strategic approach that combines a top-down mandate with bottom-up capacity building.

- **Conduct a Diagnostic Assessment:** The first step is to conduct a diagnostic assessment to understand the existing M&E capacity and needs of each MDA. This helps to identify gaps in skills, resources, and systems, which can then be addressed through a tailored action plan.
- **Embed a Dedicated Budget:** A crucial best practice is to ensure that M&E units have a specific, non-negotiable budget line item within the MDA's annual budget. This provides the unit with the financial resources to conduct its work and signals that M&E is a core function.
- **Phased Implementation and Champions:** Instead of a full-scale, simultaneous rollout, a phased implementation approach can be more effective. Start with a few pilot MDAs that have strong leadership buy-in and a clear need for M&E. These successful examples can then serve as champions for other ministries.
- **Link M&E to Performance Management:** Making M&E a part of the performance management and appraisal systems for senior managers and program officers creates a powerful incentive for compliance and evidence use.
- **Clear Roles and Responsibilities:** The roles of the M&E unit must be clearly defined in relation to other departments within the ministry (e.g., planning, finance) to avoid duplication of effort and to ensure that the M&E unit's mandate is respected.

In summary, the development of M&E units within MDAs is a complex but essential process. These units are the operational heart of the national M&E system, responsible for implementing the national framework and providing the critical data needed for accountability and learning. By addressing the common challenges through a strategic and well-resourced approach, governments can ensure that M&E is not just a top-down mandate but a bottom-up, evidence-based culture of performance.

5.4.1 Inter-governmental and Multi-level M&E Systems

In many African countries, governance is structured across multiple tiers, from a national central government to regional provinces or states, and down to local municipalities or districts. In such a decentralized system, the institutionalization of M&E must also be multi-level. A single, national-level M&E system is insufficient to ensure accountability and performance when the vast majority of public services are delivered at the sub-national level. This section provides a comprehensive discussion on the importance, structure, and challenges of developing inter-governmental and multi-level M&E systems, drawing on practical lessons from countries with decentralized governance structures.

5.4.2 The Rationale for a Multi-level M&E System

The need for a multi-level M&E system is a direct consequence of decentralized governance. In a decentralized state, local governments are not simply administrative extensions of the central government; they are autonomous entities with their own mandates, budgets, and accountability to their local constituencies. Therefore, an effective M&E system must be designed to function at all levels for several key reasons:

- **Accountability for Service Delivery:** Local governments are the primary providers of essential services such as water, sanitation, local infrastructure, and primary healthcare. For a government to be truly accountable for its performance, M&E must be conducted at the local level to track the quality, quantity, and equity of these services.

- **Informing Local Decision-Making:** Local governments require M&E data to make informed decisions about resource allocation and program management in their specific context. An M&E unit in a municipality, for example, needs data on local roads to prioritize which ones to maintain or repair.
- **A Coherent National Picture:** The central government needs to understand if its national policies and programs are being effectively implemented at the local level. A multi-level M&E system provides the necessary data flow to aggregate local performance information into a coherent national picture for strategic planning, reporting to Parliament, and for international reporting obligations.

5.4.3 The Structure and Roles in a Multi-level M&E System

A well-designed multi-level M&E system is a "system of systems," where each level of government has a distinct but interconnected role.

- **The National Level: The Architect and Steward:** The central government's role is to act as the architect and steward of the entire system. This involves:
    - **Policy and Frameworks:** Developing a national M&E policy and a comprehensive framework that provides common standards, methodologies, and reporting protocols for all levels of government.
    - **Coordination and Oversight:** The central M&E body (e.g., the Presidency or Ministry of Planning) is responsible for coordinating the M&E efforts of all provinces and municipalities, ensuring that data is submitted on time and is of high quality.
    - **National Reporting:** Aggregating data from all levels to produce a national performance report, which is then used for high-level decision-making and accountability.

- **The Provincial/State Level: The Bridging Function:** In a multi-level system, the provincial or state level plays a crucial bridging role. It is the key link between the national government's broad mandate and the local governments' operational realities. The provincial M&E function is responsible for:
    - **Sub-national Coordination:** Coordinating the M&E activities of the provincial ministries and providing support to local governments within their jurisdiction.
    - **Data Aggregation and Reporting:** Aggregating performance data from local governments and reporting it upwards to the national level, as well as providing feedback and technical support to the local level.
- **The Local Government Level: The Implementer:** The local government, such as a municipality or a district, is where M&E becomes an operational reality. At this level, M&E units are responsible for:
    - **Program Monitoring:** The day-to-day monitoring of local projects and programs, tracking progress, and ensuring resources are being used effectively.
    - **Data Collection:** Collecting raw data on service delivery and local interventions, often from the front lines of service provision.
    - **Local Reporting:** Preparing and submitting performance reports to their own local council, as well as upwards to the provincial and national levels. In **South Africa**, for example, the **Municipal Systems Act (MSA)** provides a legal mandate for local governments to implement performance management systems with key performance indicators to measure their developmental interventions.

5.4.4. Challenges of Multi-level M&E Systems

Building a functional multi-level M&E system is fraught with significant challenges that require careful navigation.

- **Capacity Gaps:** M&E capacity is often weakest at the local government level. Municipalities often lack the financial resources and technically skilled human personnel to collect data, analyse it, and produce quality reports. This can lead to a breakdown in the entire system, as the quality of national reporting is only as good as the data collected at the lowest level. A study on South African municipalities consistently highlighted this challenge, pointing to "inadequate resources" and "low capacity" as major hindrances.
- **Data Inconsistencies and Reporting Fatigue:** A major challenge is ensuring that data collected at different levels is consistent and comparable. Without a standardized framework, data from one municipality may not be comparable to another. This can be compounded by "reporting fatigue," where different levels of government request similar information, creating a significant administrative burden on local governments.
- **Political and Administrative Barriers:** The relationship between different levels of government can be complex, marked by political differences and administrative "turf battles." A lack of political will or cooperation from a provincial or local government can derail the entire system, as the central government may lack the authority to enforce its M&E mandate.
- **Lack of Incentives:** Local governments may not see a direct benefit in investing time and resources in M&E, particularly if the M&E reports are primarily used for upward accountability to the central government. Without clear

incentives for performance, M&E can become a compliance exercise rather than a tool for learning and improvement.

5.4.5. Best Practices for Overcoming Challenges

Despite the challenges, countries like **South Africa** and **Uganda** have made progress in navigating the complexities of multi-level M&E. Their experiences offer several best practices:

- **A Coherent Legal and Policy Framework:** A clear legal and policy framework is essential. The **South African** model, which has both a national M&E policy and a legal mandate at the municipal level (via the MSA), provides a strong foundation.
- **Targeted Capacity Building:** The central government should provide targeted and sustained capacity-building support to sub-national governments. This includes training programs, toolkits, and dedicated technical assistance to help local M&E units build their skills.
- **Integrated Reporting Systems:** The national government should develop integrated reporting systems that minimize reporting burdens on sub-national governments. The goal should be to design a system where data collected for local use can be easily aggregated for national reporting.
- **Fostering a Culture of Mutual Accountability:** The relationship between different levels of government should be built on a culture of mutual accountability and support. The national government should not just demand reports but also use M&E findings to provide constructive feedback, share best practices, and offer additional support to struggling sub-national entities.
- **Local-Level M&E Champions:** The central government should work to identify and empower M&E "champions" at

the local level. These are individuals and leaders who understand the value of M&E and can drive its implementation within their own government.

In summary, a multi-level M&E system is a necessity for effective governance in a decentralized state. It is the mechanism that connects national policy to local service delivery, ensures that a government can be held accountable from the top down, and empowers local leaders to make evidence-based decisions. While its implementation is challenging, it is a crucial component of institutionalizing M&E and building a more responsive, transparent, and performance-oriented public sector.

Chapter 6

Capacity Building and Human Resources

Introduction

The establishment of a robust Monitoring and Evaluation (M&E) system, with its legal frameworks, central coordinating bodies, and ministry-level units, is a significant achievement. However, even the most perfectly designed system will fail without the human skills, institutional support, and enabling environment to make it function. M&E capacity is the engine that drives the entire system, and a thorough assessment of these capacity needs is the critical first step toward building a functional and sustainable M&E culture. This chapter provides a comprehensive discussion on the imperative of capacity assessment, the three levels of M&E capacity, and practical methodologies for moving from assessment to a targeted capacity-building strategy.

6.1.1 The Imperative for Capacity Assessment

A government's commitment to M&E often begins with a broad mandate, but the subsequent capacity-building efforts can be a shot in the dark if they are not based on a clear understanding of existing strengths and weaknesses. A comprehensive M&E capacity assessment serves as a diagnostic tool, providing an evidence-based roadmap for all subsequent interventions. The imperative for such an assessment is rooted in several key benefits:

- **Targeted Interventions:** An assessment identifies specific skill gaps, resource deficiencies, and institutional bottlenecks, allowing a government to design a targeted capacity-building strategy rather than a generic, one-size-fits-all training program.
- **Effective Resource Allocation:** By pinpointing where capacity is weakest, an assessment ensures that limited financial resources are allocated to the most critical areas,

maximizing the impact of capacity-building investments.

- **Building a Shared Vision:** The assessment process itself, which often involves broad stakeholder consultation, helps to build a shared understanding of M&E's purpose and fosters a sense of ownership among practitioners and senior leaders.
- **Establishing a Baseline:** A baseline assessment provides a benchmark against which future progress can be measured, allowing a government to monitor the effectiveness of its capacity-building efforts over time.

6.1.2 The Three Levels of M&E Capacity

M&E capacity is not a single, monolithic concept. It exists at three interconnected levels, and an effective assessment must look at all three to understand the complete picture.

1. Individual Level Capacity: The Human Skills

This level of capacity refers to the knowledge, skills, and attitudes of the individuals responsible for M&E functions. A strong M&E system is built on the expertise of its people. An assessment at this level would seek to understand:

- **Technical Skills:** The ability to develop logical frameworks, design data collection tools, manage databases, and conduct data analysis using both quantitative and qualitative methods.
- **Evaluation Competencies:** The skills required to manage or conduct evaluations, including developing terms of reference, selecting consultants, and assessing the quality of evaluation reports.
- **Soft Skills and Attitudes:** The ability to communicate M&E findings to different audiences, work collaboratively with program managers, and champion a culture of evidence use.

2. Organizational Level Capacity: The Institutional Structures

This level of capacity refers to the institutional frameworks, processes, and resources that enable and support the M&E function within an organization, such as a ministry or a department. An assessment at this level would examine:

- **Structures and Roles:** The existence of a dedicated M&E unit or a clearly defined M&E focal person, with a clear mandate and reporting lines.
- **Resources:** The availability of a dedicated budget for M&E activities, as well as access to the necessary IT infrastructure and data management tools.
- **Processes and Frameworks:** The existence of a clear M&E plan for the organization, standardized reporting templates, and a documented system for how M&E findings are used in decision-making. The case study of the South African Department of Home Affairs (DHA), which has a "well-constituted M&E Directorate," demonstrates a strong organizational capacity that is key to enhancing service delivery.

3. Systemic Level Capacity: The Enabling Environment

This is the broadest level, referring to the legal, political, and cultural context in which M&E operates. It is the enabling environment that determines whether individual and organizational capacities can be effectively leveraged for change. An assessment at this level would look at:

- **Legal and Policy Framework:** The existence of a national M&E policy, a legal mandate, and the presence of sanctions or incentives for compliance.
- **Political Will and Leadership:** The degree to which senior political and administrative leaders demand and use M&E

information for decision-making.

- **Accountability and Data Ecosystem:** The existence of a vibrant community of M&E practitioners, a robust national data ecosystem, and the role of parliamentary oversight committees in using M&E data for public accountability. A study on Botswana's M&E system, for example, highlighted the systemic failure of an M&E system to link to "core functions" and the broader "contextualisation issues," which is a clear example of systemic capacity challenges.

6.1.3 Methodologies for M&E Capacity Assessment

A comprehensive capacity assessment typically employs a mixed-methods approach to gather data from multiple sources. A popular tool, the Monitoring and Evaluation Capacity Assessment Toolkit (MECAT), used in countries like Kenya, provides a clear methodology. The following methods are commonly used:

- **Document Review (Desk Review):** Analysing existing M&E policies, strategic plans, reporting templates, and M&E reports to understand the existing formal structures and processes. This provides a clear picture of the "de jure" system.
- **Key Informant Interviews:** Conducting semi-structured interviews with a cross-section of stakeholders, including senior management, M&E practitioners, and central M&E body staff. This provides qualitative insights into the "de facto" reality—the challenges and successes that are not captured in official documents.
- **Surveys and Questionnaires:** Administering a structured survey to M&E practitioners across government to collect quantitative data on their skills, resources, and perceptions of the M&E system.
- **Participatory Group Assessments:** Conducting focus group

discussions with small groups of M&E staff or program managers to jointly diagnose capacity gaps and build a shared understanding of the problems. The participatory nature of this method helps to build ownership of the assessment findings.

6.1.4 From Assessment to Action: Designing a Capacity-Building Strategy

The ultimate goal of a capacity assessment is not just to identify problems but to provide a clear path forward. The findings of the assessment should be used to design a targeted capacity-building strategy with interventions at all three levels.

- **Individual-Level Interventions:**
  - **Training and Mentorship:** Delivering specialized training on M&E methodologies, data analysis software, and reporting skills.
  - **On-the-Job Support:** Providing mentorship and coaching to M&E focal points to help them apply their new skills in their day-to-day work.
  - **Communities of Practice:** Fostering professional networks and communities of practice where M&E practitioners can share knowledge and support one another.
- **Organizational-Level Interventions:**
  - **Institutional Strengthening:** Supporting ministries in establishing dedicated M&E units, defining clear job descriptions, and developing M&E plans that are integrated into their core functions.
  - **Resource Mobilization:** Advocating for the allocation of a dedicated budget for M&E activities and helping ministries acquire the necessary hardware and software.
  - **Process Improvement:** Streamlining data collection and

reporting processes to reduce administrative burden and improve the quality of information flow.

- **Systemic-Level Interventions:**
    - **Policy and Legal Advocacy:** Using the assessment findings to advocate for strengthening the national M&E policy, clarifying legal mandates, and establishing a robust system of incentives and sanctions.
    - **Promoting an M&E Culture:** Hosting high-level forums, publicizing M&E success stories, and engaging with parliamentary oversight committees to foster a national culture where evidence-based decision-making is valued and rewarded.

Assessing M&E capacity needs is a non-negotiable step toward building a sustainable M&E system. By systematically diagnosing strengths and weaknesses at the individual, organizational, and systemic levels, a government can move beyond a reactive, ad hoc approach to M&E and design a targeted, strategic, and effective capacity-building program. The ultimate aim is to ensure that the entire government, from the President's office to the village council, has the skills, resources, and enabling environment to use evidence to drive performance and deliver on its promises to the people.

6.2 Strategies for Training and Professional Development for Civil Servants

Following a thorough M&E capacity needs assessment, the next critical step is to design and implement a strategic and targeted capacity-building program. The success of any M&E system, no matter how well-designed, hinges on the skills, knowledge, and attitudes of the civil servants who must operate it. This section provides a comprehensive discussion on effective strategies for training and professional development for civil servants, moving beyond a one-off workshop to a holistic and sustained approach that fosters a genuine culture of M&E.

6.2.1 The Strategic Imperative of M&E Training

In a results-oriented public sector, M&E training is not a luxury but a fundamental necessity. Its strategic importance lies in its ability to:

- **Build Technical Competence:** Training equips civil servants with the technical skills to design, implement, and manage M&E systems, from developing logical frameworks to conducting data analysis and writing evidence-based reports.
- **Cultivate an M&E Culture:** The ultimate goal of training is to shift mindsets. It aims to move civil servants from seeing M&E as a bureaucratic burden to viewing it as an indispensable tool for organizational learning, improved performance, and enhanced accountability.
- **Improve Service Delivery:** By enabling civil servants to better monitor and evaluate their programs, training directly contributes to more effective and efficient service delivery. When managers can use data to identify what works and what doesn't, they can make better decisions that benefit citizens.

6.2.2 Core Principles of Effective M&E Training

For M&E training to be effective, it must be guided by a set of core principles that move it from a theoretical exercise to a practical, impactful endeavour.

- **Targeted and Tailored:** Training must not be a one-size-fits-all program. It must be tailored to different audiences and their specific roles. For instance, a senior manager requires training on how to use M&E findings for strategic decision-making, while an M&E practitioner needs hands-on training in data collection tools and analysis software. A program offered by the National School of Government (NSG) in

South Africa exemplifies this, with courses targeted at "M&E practitioners" and "M&E managers" with distinct learning outcomes for each group.

- **Practical and Hands-on:** Effective M&E training is not about passive listening. It must be practical and hands-on, using real-world case studies and data from the trainees' own organizations. This ensures that civil servants can immediately apply what they learn to their day-to-day work, which reinforces the learning and demonstrates the value of M&E.
- **Blended Learning Approach:** A blended learning approach that combines various training modalities is more effective and sustainable. This can include traditional classroom workshops, online e-learning modules, and structured on-the-job learning through mentorship and coaching. This approach caters to different learning styles and ensures that learning continues long after a formal training course has ended.

6.2.3 A Multi-faceted Approach to Professional Development

Professional development for M&E should be seen as a continuous process, not a one-time event. An effective strategy uses a multi-faceted approach that addresses the needs of civil servants at different stages of their careers and within different roles.

- **Formal Training and Certification:** This is the most structured form of professional development. Governments can partner with national training institutions, universities, and professional associations to offer accredited courses and certifications. These programs, which can range from short courses to postgraduate diplomas, provide a strong theoretical foundation and a common language for M&E across the public service. For instance, a course offered by the NSG in South Africa provides a structured curriculum that covers "use of indicators," "quantitative research methods," and "report writing," with a clear assessment and a portfolio of evidence to demonstrate competence.

- **On-the-Job Learning and Mentorship:** This is arguably the most effective strategy for embedding M&E skills. It involves a "learning-by-doing" approach where civil servants are coached by experienced M&E professionals within their own ministries. A notable example is the Track20 program in Kenya, where M&E officers were embedded directly into the Ministry of Health to work on and improve the government's data systems. This approach not only builds individual capacity but also helps to strengthen organizational systems and create a lasting M&E culture from within.
- **Communities of Practice and Peer Learning:** Creating a platform for M&E practitioners to connect, share experiences, and solve problems is a powerful strategy for professional development. These communities of practice can be virtual or in-person forums where civil servants from different ministries can discuss common challenges, share best practices, and develop a common language on M&E. This also helps to break down "silo mentalities" and fosters a culture of mutual support.
- **Academic Partnerships:** For more advanced professional development, governments can form partnerships with universities and research institutions to offer advanced degrees or research opportunities in M&E. This not only builds a pool of highly skilled M&E professionals but also creates a bridge between academia and the public sector, ensuring that M&E practices remain current and evidence-based.

6.2.4 Case Studies: M&E Training in the African Context

Several African countries have implemented innovative strategies for M&E professional development, providing valuable lessons for others.

- **South Africa:** The institutionalization of M&E in South Africa is underpinned by a robust training and capacity development framework. The Department of Planning, Monitoring, and Evaluation (DPME) works with institutions like the National School of Government and various universities to provide a wide range of courses. These programs are specifically designed to address the challenges of the South African public sector, such as linking M&E to the planning system and using evidence to improve performance. This multi-institutional approach ensures a sustained supply of M&E skills and a common understanding of national frameworks.
- **Kenya:** The Track20 program, in collaboration with the government, provides a unique example of a demand-driven approach. By embedding M&E professionals directly into ministries, the program was able to provide hands-on mentorship and technical support, directly addressing the ministry's specific M&E challenges and building lasting capacity. This model demonstrates the effectiveness of moving beyond a purely classroom-based approach to a more practical, on-the-job mentorship model.

6.2.5 Challenges and Recommendations

Despite the clear benefits, M&E training and professional development face several common challenges.

- **High Staff Turnover:** A persistent challenge is the loss of institutional memory when trained staff are transferred or leave the public service.

- **Lack of Dedicated Budgets:** Ministries often lack a dedicated budget for professional development, viewing it as an expense rather than an investment.
- **Resistance and Low Motivation:** Some civil servants may be resistant to new M&E practices, particularly if they see it as an additional workload without clear incentives.

To overcome these challenges, governments should adopt the following recommendations:

- **Link Training to Career Progression:** Make M&E skills a prerequisite for promotion within the civil service. This creates a powerful incentive for civil servants to seek out training and to apply their skills in their work.
- **Embed a Long-Term Professional Development Framework:** Move away from one-off training events and embed M&E training into a long-term professional development framework that is tied to performance appraisals.
- **Champion M&E from the Top:** Senior political and administrative leaders must champion M&E and actively use M&E findings in their decision-making. This sends a clear signal to all civil servants that M&E is valued and not just a tick-box exercise.

M&E training and professional development are not just about building individual skills; they are about institutionalizing a culture of accountability, learning, and performance. By adopting a strategic, multi-faceted, and sustained approach, governments can empower their civil servants to become effective managers of development, capable of delivering better results for their citizens.

6.3.1 The Role of Universities and Training Institutes in Building a Professional M&E Workforce

While governments are ultimately responsible for institutionalizing M&E, they cannot do it alone. The demand for a skilled and professional M&E workforce far outstrips the capacity of any single public sector institution to produce it. This is where universities, training institutes, and other academic bodies play a crucial and indispensable role. These external partners constitute the "supply side" of M&E capacity, building the foundational knowledge, advanced skills, and professional identity that are essential for a credible and effective M&E system. This section provides a comprehensive discussion on the distinct roles of these institutions, the importance of a collaborative ecosystem, and the challenges and best practices for forging an effective partnership between government and academia.

6.3.2 The Strategic Importance of External Partners

A government's M&E system is only as strong as the people who operate it. Universities and training institutes provide a level of expertise, independence, and long-term vision that is difficult for a government to replicate. Their strategic importance lies in their ability to:

- **Professionalize the Field:** By offering accredited degrees and diplomas, these institutions elevate M&E from a collection of ad-hoc tasks to a recognized profession with its own body of knowledge, methodologies, and ethical standards.
- **Ensure Independence and Credibility:** External institutions provide a source of independent thought and expertise, which is crucial for conducting credible evaluations and challenging assumptions within the public sector. This independence is a

cornerstone of a robust and trustworthy M&E system.

- **Drive Research and Innovation:** Universities are centres for research and intellectual leadership. They are at the forefront of developing new M&E methodologies, adapting global best practices to local contexts, and providing the academic rigor that strengthens the entire field.
- **Build a Sustainable Talent Pipeline:** By offering formal education and training, these institutions are responsible for creating a steady pipeline of new talent, ensuring a continuous supply of skilled M&E practitioners for the public and private sectors.

6.3.3 The Contribution of Universities and Academic Institutions

Universities play a distinct role in building a professional M&E workforce, focusing on long-term education and the development of a strong academic foundation. Their contributions include:

- **Offering Formal Degrees and Diplomas:** Universities are the primary institutions for offering formal qualifications in M&E. These programs, which can range from postgraduate diplomas to master's degrees, provide a comprehensive education that goes beyond technical skills to include a deep understanding of governance, public policy, and research methodologies. For instance, the University of the Witwatersrand (Wits) in South Africa offers a "Postgraduate Diploma in Public Management in the field of Monitoring and Evaluation," which is specifically designed for working professionals to enhance their skills in a results-oriented public sector. Similarly, Stellenbosch University and the University of Cape Town offer master's programs in M&E, demonstrating a strong academic ecosystem. In Zimbabwe, we have Lupane State University and Midland State University offering postgraduate and master's degrees in M&E.

- **Conducting Research and Publishing:** Universities are vital for generating the intellectual capital that underpins the M&E field. Their research helps to develop and refine M&E theories and methodologies, and their publications provide a public repository of knowledge that can be accessed by practitioners and policymakers. This research is also crucial for adapting global M&E standards to the specific realities and contexts of African countries.
- **Serving as Hubs for Knowledge and Dialogue:** Universities can host conferences, seminars, and workshops that bring together M&E professionals, policymakers, and civil society leaders. These forums are crucial for fostering a vibrant M&E community and promoting dialogue on key issues.

6.3.4 The Contribution of Training Institutes and Vocational Centres

While universities focus on long-term professionalization, training institutes and vocational centres fill a critical and complementary role by providing short-term, practical, and highly targeted training.

- **Targeted and Practical Skills Transfer:** These institutions are highly agile and can design courses that respond quickly to the immediate needs of civil servants. They excel at providing practical, hands-on training in specific M&E tools and techniques, such as a course on designing a logframe, a workshop on data visualization, or training on a specific M&E software. The National School of Government (NSG) in South Africa, which is a government-mandated institution, is a perfect example of this. It offers a wide range of training programs that are designed to build the M&E capacity of civil servants at all levels, from entry-level to senior management.

- **Bridging Theory and Practice:** Training institutes are uniquely positioned to bridge the gap between the academic theories taught in universities and the practical realities of government work. They can create courses that are directly relevant to the daily challenges faced by civil servants, helping them apply M&E principles to their specific programs and projects.

6.3.5 A Collaborative Ecosystem: Case Studies

The most effective approach to building a professional M&E workforce is not for the government to work with a single institution, but to foster a collaborative ecosystem of universities, training institutes, and professional associations.

- **South Africa:** The South African system provides a strong example of such an ecosystem. The government's M&E mandate is supported by a multi-pronged approach that includes the National School of Government (NSG) for vocational training, universities like WITS for accredited degrees, and the South African Monitoring and Evaluation Association (SAMEA), which provides a professional home for M&E practitioners and offers internships for emerging evaluators. This collaborative model ensures that the demand for M&E skills is met with a high-quality supply from various sources.
- **Regional Initiatives:** The Centre for Learning on Evaluation and Results (CLEAR), which is a global initiative with a regional centre for Anglophone Africa, is a prime example of a multi-stakeholder partnership. Housed at a university,

CLEAR works to strengthen M&E systems across multiple countries by providing training, technical assistance, and a platform for regional knowledge sharing.

6.3.6 Challenges and Recommendations for an Effective Partnership

Despite the clear benefits of this partnership, several challenges can hinder its effectiveness.

- **Mismatch Between Theory and Practice:** There can sometimes be a disconnect between the academic theories taught in universities and the practical realities of public sector work.
- **Lack of Sustainable Funding:** Many public servants face a lack of dedicated funding for professional development, making it difficult to enrol in university programs or even attend short training courses.
- **Weak Communication and Collaboration:** A lack of regular communication and collaboration between the government, universities, and training institutes can lead to a mismatch between the skills being taught and the skills that are actually needed in the public sector.

To overcome these challenges, governments and their partners should:

- **Formalize Partnerships:** Establish formal, long-term partnerships with universities and training institutes to ensure a consistent and high-quality supply of M&E education.
- **Create a Professional Development Framework:** Develop a clear professional development framework for the civil service that specifies what M&E skills are needed at each career level

and which courses or certifications are required.

- **Bridge the Gap:** Encourage a more practical approach to university education by incorporating real-world case studies, internships, and guest lectures from government practitioners.
- **Fund Professional Development:** Dedicate a specific budget line for the professional development of M&E staff and offer scholarships or other forms of financial support for civil servants to pursue M&E qualifications.

Universities and training institutes are critical partners in a government's journey to institutionalize M&E. By building a professional workforce and fostering a culture of evidence, they provide the intellectual, academic, and practical foundations for a more accountable, transparent, and results-oriented public sector.

6.4.1 Addressing Staff Retention and Creating Career Paths for M&E Specialists

The journey to institutionalizing a robust M&E system is a significant and long-term investment. Governments dedicate considerable resources to establishing legal frameworks, creating organizational structures, and building the capacity of their civil servants through training and professional development. However, these investments can be completely undermined if the trained and skilled M&E professionals do not stay in the public service. Staff retention is the final, and perhaps most critical, piece of the M&E capacity puzzle. This section provides a comprehensive discussion on the challenges of retaining M&E specialists and outlines a set of strategic interventions for creating clear career paths and a sustainable M&E workforce within the public sector.

6.4.2 The Challenge of Staff Retention in the Public Sector

High turnover among M&E specialists is a pervasive problem in many public sectors. The reasons for this are multi-faceted, and they often lead to a significant loss of institutional memory and a continuous need to re-train new staff. Some of the key challenges include:

- **Unclear Career Paths:** In many government ministries, M&E is not a recognized professional discipline with a dedicated career ladder. Staff are often "assigned" M&E tasks in addition to their core duties, with no clear path for progression. This lack of a formalized career trajectory is a major demotivator for ambitious professionals who want to grow in their field.
- **Uncompetitive Compensation:** The M&E sector is a global one, with a high demand for skilled professionals from NGOs, international development organizations, and the private sector. These organizations often offer significantly higher salaries and benefits, making it difficult for the public sector to compete.
- **Lack of Professional Identity and Recognition:** Without a clear scheme of service or a professional cadre, M&E specialists can feel isolated and undervalued. Their work may not be formally recognized in performance appraisals or promotion criteria, and they may be perceived as "support staff" rather than as a core part of the management team.
- **Limited Opportunities for Professional Growth:** Beyond basic training, many governments do not provide a clear pathway for continuous professional development. This lack of funding for advanced degrees, certifications, or participation in professional forums and conferences can stifle a specialist's growth and push them to seek opportunities elsewhere.
- **Lack of Leadership Buy-in:** When senior leaders do not actively use M&E findings in their decision-making, the M&E function can be seen as a bureaucratic exercise. This lack

of visible impact can make M&E professionals feel that their work is not valued and that their efforts are in vain.

6.4.3 Strategies for Creating a Sustainable M&E Workforce

Addressing staff retention requires a strategic, multi-pronged approach that goes beyond simply increasing salaries. It involves institutionalizing M&E as a professional career path and building a culture that values the M&E function.

- **Formalizing a Professional M&E Cadre:** This is the single most important step a government can take. The creation of a dedicated M&E cadre or a "scheme of service" formalizes M&E as a legitimate profession within the civil service. A scheme of service is a legal document that clearly defines the qualifications, duties, and responsibilities for each grade level, from entry-level M&E Officer to senior M&E Director. It provides a clear and transparent career ladder, showing civil servants that a career in M&E is a viable and respected long-term option. For example, a senior M&E role in an organization like the **Africa CDC** has clear requirements, such as a master's degree and a minimum of 10 years of experience, providing a tangible example of a formalized career path.
- **Providing Continuous Professional Development:** Governments must invest in the long-term growth of their M&E specialists. This includes providing a budget for:
    - **Advanced Education:** Sponsoring civil servants to pursue postgraduate diplomas or master's degrees in M&E at universities.
    - **Certifications and Short Courses:** Funding participation in specialized training on new methodologies, data analysis software, or qualitative research.

    - **Professional Engagement:** Supporting attendance at national and international M&E conferences, workshops, and forums, which provide valuable networking and learning opportunities.
- **Reviewing and Adjusting Remuneration:** While salary is not the only motivator, it is a significant one. Governments must review the remuneration and benefits packages for M&E professionals and make them more competitive with the market. This does not mean matching the private sector one-for-one, but rather ensuring that public sector M&E specialists are compensated in a way that reflects their technical skills, educational qualifications, and the strategic importance of their work.
- **Building a Community of Practice:** A sense of professional identity and mutual support is a powerful retention tool. Governments can foster a community of M&E practitioners through the creation of a public sector-wide M&E forum or by supporting a national Voluntary Organisation for Professional Evaluation (VOPE). As seen in Ghana, a strong VOPE can play a crucial role in advocacy, knowledge sharing, and professional development, providing a sense of belonging and purpose for M&E specialists.
- **Championing the Use of M&E Findings:** The work of M&E professionals must be visibly valued by senior leadership. When leaders use M&E findings to inform policy decisions, re-allocate budgets, or improve program design, it sends a clear message that M&E is not a bureaucratic exercise but a strategic function. This public recognition is a powerful, non-monetary incentive that motivates professionals and makes them feel that their work is impactful.

6.4.4 Case Studies from the African Public Sector

Several African countries have made significant strides in professionalizing the M&E function, providing valuable case studies for others to learn from.

- **South Africa:** The establishment of the Department of Performance Monitoring and Evaluation (DPME) and the work of the Public Service Commission (PSC) are a testament to the country's commitment to institutionalizing M&E. The PSC has long been involved in the "professionalizing of the Public Service," and its discussions on "personnel practices" and creating a "common language on M&E" are foundational to building a professional M&E workforce. While a dedicated cadre may not exist in all ministries, the existence of a high-level department and supportive institutions sends a clear signal about the value of M&E as a career.
- **Mauritius:** The Mauritian Civil Service provides a clear example of the formalization of career paths through its "Scheme of Service" for various cadres. While not exclusive to M&E, this framework provides a model for how a government can legally define qualifications, responsibilities, and progression for a specific group of professionals. A similar scheme could be created for M&E specialists to formalize their role and career trajectory.

6.4.5 Recommendations for Action

Based on these challenges and strategies, governments committed to building a sustainable M&E workforce should:

- **Formalize a Dedicated M&E Cadre:** Work with the Ministry of Public Service to create a formalized M&E cadre or scheme of service that clearly defines a career ladder from entry-level to senior positions, complete with standardized job descriptions and promotion criteria.

- **Allocate a Specific Budget for Professional Development:** Dedicate a budget line for the continuous professional development of M&E specialists, including funding for accredited courses, certifications, and conference attendance.
- **Create Incentives for High-Quality Work:** Link M&E work to performance appraisals and reward staff for high-quality M&E reports, innovative use of data, or successful evaluations.
- **Promote an M&E-Friendly Culture:** Conduct regular high-level forums where senior leaders publicly discuss and use M&E findings, thereby demonstrating the value and importance of the M&E function to the entire civil service.

In conclusion, a government's investment in M&E capacity is only as valuable as its ability to retain the skilled professionals who are trained. By strategically addressing the challenges of staff retention through formalized career paths, continuous professional development, and a culture that values the M&E function, governments can build a sustainable and professional M&E workforce that is capable of driving a results-oriented public sector.

Chapter 7

Data Management and Information Systems

Introduction

The journey of institutionalizing a national M&E system begins with establishing the legal and organizational frameworks, building human capacity, and fostering a culture of accountability. However, in the 21st century, the effectiveness of any M&E system is ultimately determined by its technological backbone: a robust and integrated M&E information system. Without a centralized system for data management, the information collected by various government ministries and agencies remains in fragmented silos, making it impossible to get a coherent national picture of development progress. This chapter provides a comprehensive discussion on the strategic imperative of a national M&E information system, its key components, a best-practice approach to implementation, and a look at real-world case studies from the African public sector.

7.1.1 The Imperative of a National M&E Information System

In a modern, results-oriented public sector, a national M&E information system is not a luxury; it is a fundamental requirement for effective governance. Its strategic importance lies in its ability to:

- **Establish a Single Source of Truth:** A national system consolidates all M&E data into a single, authoritative repository. This eliminates the problem of conflicting data from different ministries and ensures that policymakers, the public, and development partners are all working from the same facts.
- **Improve Data Quality and Integrity:** By standardizing reporting templates, implementing data validation rules, and providing a clear data collection hierarchy, an information system significantly improves the quality, accuracy, and reliability of M&E data.

- **Enable Real-time Performance Monitoring:** Manual, paper-based reporting systems are slow and inefficient. A modern information system allows for the real-time collection, analysis, and visualization of data, enabling policymakers to quickly identify bottlenecks and make timely adjustments to programs.
- **Facilitate Cross-sectoral Analysis:** National development plans are rarely confined to a single ministry. An integrated M&E system allows for the cross-referencing and analysis of data across different sectors, providing insights into complex interdependencies and ensuring a whole-of-government approach to development.
- **Enhance Transparency and Accountability:** A well-designed system can include a public-facing portal or dashboard that provides citizens with access to key performance data. This promotes transparency, empowers citizens to hold their government accountable, and builds public trust in the M&E process.

7.1.2 Key Components of a National M&E Information System

A comprehensive national M&E information system is a sophisticated piece of technology composed of several interconnected modules. While the specific design may vary, a best-practice system typically includes the following core components:

- **The Results Framework Module:** This is the foundation of the system. It houses the national development plan, strategic frameworks, and the logical frameworks of individual programs and projects. It defines the objectives, outcomes, outputs, and the indicators that will be used to measure progress.
- **The Data Collection and Entry Module:** This module is the

gateway for data into the system. It should be user-friendly and support various data collection methods, including:

- ◦ **Web Forms:** For routine reporting from ministry-level M&E units.
- ◦ **Mobile Applications:** For field-level data collection by monitors and project officers, enabling the use of GPS-tagged photos and offline data entry.
- ◦ **API Integration:** To allow for the automated transfer of data from other government information systems, such as a health management information system or a treasury's financial management system.

- **The Data Management and Storage Module:** This is the secure, centralized database that stores all M&E data. It is crucial for ensuring data integrity, security, and a single source of truth. This module includes features for data validation, version control, and user access rights to ensure that only authorized personnel can enter or modify data.
- **The Data Analysis and Visualization Module:** This is where raw data is transformed into actionable intelligence. This module includes:
    - ◦ **Interactive Dashboards:** Providing a graphical overview of key performance indicators (KPIs) at national, sectoral, and sub-national levels.
    - ◦ **Ad-hoc Query Tools:** Allowing users to generate custom reports and conduct their own analysis.
    - ◦ **Geographic Information Systems (GIS) Integration:** To visualize data on maps, linking performance data to specific locations, such as schools, clinics, or project sites.
- **The Reporting Module:** This module automates the generation of standardized reports, such as quarterly performance reports, annual reviews, and progress reports against national and international goals (e.g., Sustainable

Development Goals). This saves significant time and effort, as the reports are generated from a single source of truth with a single click.

- **The Evaluation Management Module:** This module helps to manage the entire evaluation process, from the development of terms of reference and the selection of consultants to the storage of final evaluation reports and management responses. This ensures that evaluation findings are systematically archived and used for organizational learning.

7.1.3 A Phased Approach to Implementation

Implementing a national M&E information system is a complex undertaking that requires careful planning and a phased approach. A "big bang" rollout is often risky and can lead to system failure and institutional resistance. A more effective strategy involves the following phases:

- **Phase 1: Needs Assessment and System Design:** This initial phase involves a thorough needs assessment and stakeholder consultation. The government must clearly define the objectives of the system, identify the data requirements of different users, and select a technology platform (e.g., open source vs. commercial) that is suitable for its context.
- **Phase 2: Pilot Implementation:** Before a full-scale rollout, a Minimum Viable Product (MVP) should be piloted in a small number of ministries or departments. This pilot phase is crucial for testing the system's functionality, identifying and resolving bugs, and getting user feedback in a controlled environment.
- **Phase 3: Full-Scale Rollout and Training:** Once the pilot is successful, the system can be rolled out across all government ministries and agencies. This phase must be accompanied by

a comprehensive training and change management program to ensure that civil servants are equipped with the skills and confidence to use the new system effectively.

- **Phase 4: Continuous Improvement and Support:** The implementation of the system is not the end of the journey. The final phase involves continuous technical support, system maintenance, and upgrades to ensure that the system remains relevant and responsive to the evolving needs of the government.

7.1.4 Challenges and Best Practices for Implementation

Designing and implementing a national M&E information system is fraught with challenges, but these can be overcome with a strategic approach and a focus on best practices.

Challenges:

- **High Initial Costs:** Developing and deploying a national-scale information system requires a significant upfront investment in software, hardware, and technical expertise.
- **Technical Capacity Gaps:** Many public sectors lack the in-house technical capacity to design, implement, and maintain a complex M&E information system.
- **Data Quality Issues:** A new system is only as good as the data that goes into it. The challenge is often not the technology itself but the underlying data quality and integrity of administrative data.
- **Institutional Resistance:** Civil servants who are used to manual, paper-based reporting may be resistant to adopting a new technology, especially if they are not properly trained or do not see the value in it.
- **Lack of Political Buy-in:** Without strong political leadership and commitment from the highest levels of government, a

national M&E information system is likely to fail.

Best Practices:

- **Ensure High-Level Political Buy-in:** The implementation of a national M&E information system must be a top-down, government-wide priority. As seen in South Africa, the creation of the Department of Performance Monitoring and Evaluation (DPME) with a clear mandate from the President was a key factor in the successful push for a national system.
- **Start Simple and Iterate:** A system should not be over-engineered from the start. A phased approach that begins with a Minimum Viable Product (MVP) and adds more features over time is a more manageable and sustainable strategy.
- **Invest in Technical and Human Capacity:** Governments must invest in both the technical infrastructure and the human capacity to use it. This involves hiring skilled IT professionals and providing a comprehensive training program for all civil servants who will interact with the system.
- **Design for the User:** The system must be designed with the end-user in mind. A user-friendly interface, clear instructions, and responsive technical support are crucial for promoting adoption and ensuring the system is actually used.
- **Emphasize Data Use, Not Just Collection:** The system's value must be demonstrated through its ability to generate useful reports and insights that inform decision-making. The goal is to avoid the perception that the system is simply a new tool for data collection without any tangible benefits.

7.1.5 Case Studies from the African Public Sector

Several African countries have embarked on the journey of designing and implementing national M&E information systems, providing valuable lessons for others.

- **Uganda's OPM-MIS:** The **Office of the Prime Minister (OPM)** in Uganda provides an excellent case study of a needs-driven implementation. Before the system, the OPM faced significant challenges with tracking funds, ensuring accountability, and managing a large number of development projects. They partnered with a technology provider to develop a custom Management Information System (MIS) that included a biometric-based payment mechanism for beneficiaries to prevent fraud. The system not only streamlined fund disbursement but also provided a single, real-time source of truth for all project information, leading to better visibility, improved documentation, and more effective project management.
- **South Africa's DPME System:** The Department of Performance Monitoring and Evaluation (DPME) in South Africa, established in the Presidency, has been at the forefront of using technology for M&E. The DPME developed a "Programme Performance Information reporting system" and is actively working on a "central data management and analytical system." This top-down, strategic approach, backed by strong political will, has allowed South Africa to create a comprehensive system for monitoring the performance of all government ministries against a set of national priority outcomes.
- **Zimbabwe's Chiringa Dashboard:** In a recent and ambitious initiative, the government of Zimbabwe has developed the "Chiringa Monitoring and Evaluation Dashboard." This

dashboard, developed in partnership with a local university, is being piloted with the goal of becoming the national standard for monitoring all government programs and projects. The initiative is a clear demonstration of the growing recognition that technology is a strategic imperative for good governance and a move to a more results-oriented public sector.

- **Zambia's GW-M&E/MIS:** The Zambian National Monitoring and Evaluation Policy provides a clear framework for the development of a "Government Wide Monitoring and Evaluation/Management Information System (GW-M&E/MIS)." This policy-level commitment is a crucial first step that provides the legal and institutional foundation for the eventual design and implementation of a national M&E information system.

The design and implementation of a national M&E information system is a complex but essential undertaking for any government committed to modern, results-oriented governance. By learning from the successes and challenges of others and by adopting a strategic, user-centric, and phased approach, governments can build a technological backbone that empowers them to use evidence to drive development and deliver on their promises to the people.

7.2 Data Collection Methods: Surveys, Administrative Data, and Qualitative Data

The integrity and utility of any national M&E system are entirely dependent on the quality of its data. While the previous chapter focused on the technological systems for managing this data, the real work begins with its collection. The choice of data collection method is not a mere technical decision; it is a strategic one that must be aligned with the M&E questions being asked. A robust and credible M&E system does not rely on a single method but rather uses a complementary mix of quantitative and qualitative data sources. This chapter provides a comprehensive discussion on the three main data collection methods—surveys, administrative data, and qualitative methods—and makes a strong case for a mixed-methods approach that integrates all three.

7.2.1 An Overview of Data Collection Methods

Data used for M&E can be broadly categorized into three types, each with a distinct purpose and set of characteristics:

- **Surveys (Quantitative Data):** This method involves collecting standardized, numerical data from a large sample of people. It is excellent for answering questions like "how many?" or "how much?"
- **Administrative Data:** This refers to data that is routinely collected by government ministries and agencies as a byproduct of their regular administrative functions (e.g., records on school enrolment, patient visits, or budget expenditures).
- **Qualitative Data:** This method involves collecting non-numerical, descriptive data to understand the "why" and "how" behind a phenomenon. It provides rich context and an understanding of people's experiences and perceptions.

7.2.2 Surveys: The Power of Quantitative Data

Surveys are a cornerstone of modern M&E, particularly for impact evaluations and large-scale performance monitoring. They are a systematic way of collecting standardized information from a representative sample of a population.

- **Types and Use:** Surveys can be used for a wide range of M&E activities, from baseline studies at the beginning of a project to end-of-project impact assessments. They can be administered in various ways, including face-to-face interviews, telephone surveys, and online questionnaires. For example, a national household survey can be used to measure the impact of a social protection program on poverty levels.

- **Strengths:**
    - **Generalizability:** When a survey uses a statistically sound sampling method, its findings can be generalized to the wider population, providing a high degree of confidence in the results.
    - **Statistical Analysis:** The numerical nature of survey data allows for rigorous statistical analysis, enabling the identification of trends, correlations, and causal links.
    - **Standardization:** Surveys use standardized questions and responses, ensuring that data collected across different regions or groups is directly comparable.
- **Weaknesses:**
    - **High Cost and Time:** Large-scale surveys, especially face-to-face ones, can be very expensive and time-consuming to implement, from questionnaire design to data collection and cleaning.
    - **Limited Context:** While surveys are excellent for telling you "what happened," they are often poor at explaining "why it happened." They may not capture the nuances, complexities, and personal experiences that underpin a particular outcome.
    - **Potential for Bias:** If not designed and implemented carefully, surveys can be subject to sampling bias, measurement error, and social desirability bias (where respondents give answers they think are expected).

7.2.3 Administrative Data: The Untapped Resource

Administrative data is a rich but often underutilized source of information for M&E. It is the data that governments routinely collect for the management and delivery of services.

- **Types and Use:** This includes school enrolment records from an Education Management Information System (EMIS), patient records from a Health Management Information System (HMIS), tax records, and service delivery statistics from various ministries. This data is particularly valuable for ongoing performance monitoring, as it is often available in real-time or near-real-time and provides a continuous stream of information. For example, a Ministry of Health can monitor vaccination rates by analysing administrative data from clinics and hospitals.
- **Strengths:**
    - **Efficiency and Low Cost:** Since the data is already being collected for another purpose, its use for M&E is often low-cost and time-efficient.
    - **Longitudinal Nature:** Administrative data can provide a historical record of performance over many years, allowing for the tracking of long-term trends.
    - **Comprehensive Coverage:** In many cases, administrative data covers the entire population served by a program (e.g., all students in a school system), eliminating the need for complex sampling.
- **Weaknesses:**
    - **Quality and Completeness Issues:** Administrative data can be fraught with inaccuracies, missing information, and inconsistencies. It is often not collected with M&E as a primary purpose, which can lead to data that is not perfectly aligned with an M&E framework.
    - **Lack of Context:** Like surveys, administrative data often provides little to no context. It can tell you that school enrolment has dropped, but it cannot explain the reasons behind it (e.g., new school fees, a drought forcing migration, or a lack of qualified teachers).

- Data Silos: In many governments, administrative data is stored in separate, incompatible systems, making it difficult to link and use for a comprehensive M&E purpose.

7.2.4 Qualitative Data: The Power of Context and Understanding

Qualitative data is essential for providing depth, context, and a richer understanding of the people and processes behind a project or program. It moves beyond the "what" and provides insights into the "how" and "why."

- **Methods and Use:** Common methods for collecting qualitative data include:
    - **Key Informant Interviews (KIIs):** In-depth, one-on-one interviews with people who have unique knowledge or a particular perspective on a program (e.g., a senior government official, a community leader).
    - **Focus Group Discussions (FGDs):** Group discussions with a small number of people (typically 6-10) to explore their shared experiences, perceptions, and attitudes.
    - **Case Studies:** An in-depth investigation of a specific individual, organization, or project to provide a holistic and detailed understanding of a phenomenon in its real-life context.
- **Strengths:**
    - **Explaining Causality:** Qualitative data is particularly powerful at explaining the reasons behind quantitative findings. For example, if a survey shows low program uptake, an FGD can reveal that the program's location is inconvenient or that the cultural norms of the community are a barrier.
    - **Capturing Nuance:** It can capture subtle but important details that are often missed by standardized surveys, such as the power dynamics within a community or the unintended consequences

of a policy.

  - **Rich, Evocative Narratives:** Qualitative data provides powerful, human-centred narratives that can make M&E reports more compelling and accessible to a wider audience.

- **Weaknesses:**
  - **Not Generalizable:** The findings from a few interviews or focus groups cannot be generalized to a wider population.
  - **Time-Intensive:** Data collection and analysis for qualitative methods can be time-consuming and labour-intensive.
  - **Researcher Bias:** The interpretation of qualitative data is subjective and can be influenced by the biases of the researcher.

7.2.5 A Mixed-Methods Approach: The Gold Standard

The most effective and credible M&E systems recognize that no single data collection method is sufficient on its own. The "gold standard" is a mixed-methods approach that strategically combines surveys, administrative data, and qualitative methods to create a comprehensive and holistic picture.

This approach involves using each method to complement the others. For example:

1. **Administrative data** from the Ministry of Education shows that primary school enrolment has been declining in a particular region.
2. A targeted **survey** is conducted in that region to quantify the extent of the decline and identify demographic groups most affected.
3. **Qualitative data** is then collected through focus group discussions with parents and key informant interviews with teachers and local officials to understand the underlying reasons for the decline (e.g., rising food prices forcing children to work, a new policy on school fees).

This integration of data from different sources is a powerful way to not only measure what is happening but also to understand why it is happening, which is essential for evidence-based decision-making. As a study on Ghana's local government projects demonstrated, involving communities in participatory M&E through methods like focus groups and feedback sessions can lead to a significant increase in project effectiveness and sustainability. Similarly, a study on poverty in Rwanda highlighted the need to move beyond purely quantitative indicators and integrate qualitative data to capture the subjective and relational dimensions of well-being.

A robust M&E system requires a strategic approach to data collection. By moving beyond a single method and embracing a mixed-methods approach that uses surveys to quantify, administrative data to monitor, and qualitative data to explain, governments can build a powerful evidence base that drives learning, accountability, and better development outcomes for their citizens.

7.3 Ensuring Data Quality, Integrity, and Timely Reporting

The preceding sections have emphasized the importance of robust M&E information systems and the strategic use of various data collection methods. However, the most sophisticated technological infrastructure and the most meticulously designed surveys are ultimately worthless if the data they produce is flawed or if it arrives too late to inform decision-making. Ensuring data quality, integrity, and timely reporting is the crucial, often-overlooked, bridge between data collection and data use. This section provides a comprehensive discussion on the challenges and practical strategies for making sure that the data flowing through a national M&E system is accurate, reliable, and available when it is most needed.

7.3.1 The Imperative of High-Quality and Timely Data

Data quality and timeliness are not mere technicalities; they are foundational to good governance and effective public service delivery. The consequences of poor data are far-reaching:

- **Misguided Policy Decisions:** Decisions based on incomplete or inaccurate data can lead to the misallocation of public funds, the design of ineffective programs, and a failure to address the most pressing needs of citizens.
- **Loss of Public Trust:** When official government reports contain contradictory or unreliable data, it erodes public trust in the government's ability to be transparent and accountable.
- **Ineffective Accountability:** A cornerstone of a results-oriented culture is holding public officials accountable for their performance. This is impossible without high-quality and timely data to track progress against targets.

7.3.2 Key Principles of Data Quality and Integrity

Data quality can be assessed against a set of key dimensions that determine its fitness for use. A data management system, and the processes that support it, should be designed to uphold these principles:

- **Accuracy:** Data must be correct and free from errors. For instance, a hospital record should accurately reflect the number of patients treated and the specific services they received.
- **Completeness:** All necessary data points must be present. A report on project implementation should not have missing entries for key indicators or milestones.
- **Consistency:** Data should be uniform across different sources and over time. An indicator on school enrolment should use the same definition and methodology for data collection in all provinces, and its historical data should be consistent.
- **Relevance:** The data must be relevant to the M&E question being asked. Collecting data that does not directly link to a program's objectives is a waste of time and resources.

- **Timeliness:** Data must be available when it is needed for decision-making. A quarterly performance report that arrives six months late is of little value to a minister trying to make a budget adjustment.
- **Integrity:** Data must be protected from unauthorized modification or corruption. A robust system should have audit trails and security measures to ensure data integrity from collection to reporting.

7.3.3 Challenges to Data Quality and Timely Reporting in the Public Sector

Despite the best intentions, public sector M&E systems in many African countries face significant challenges in ensuring data quality and timeliness. These challenges are often rooted in a combination of human, technical, and institutional factors.

- **Human Capacity Gaps:** As seen in a case study from Botswana, a common challenge is the shortage of trained personnel for M&E. Civil servants at the operational level may lack the skills to collect data properly, and new M&E cadres may have their roles and responsibilities poorly defined, leading to them being tasked with non-M&E duties.
- **Inadequate Infrastructure and Resources:** Poor infrastructure, such as unreliable electricity, a lack of computers, or limited internet access, can be a major barrier to timely data collection and reporting, particularly in rural areas. This was a key challenge documented in the implementation of an M&E task-shifting initiative in Botswana.
- **Lack of Institutional and Political Will:** A case study from South Africa highlighted that a lack of political will and an absence of a participatory approach during the inception of

an M&E system can render it ineffective. This often leads to a disconnect between data producers at the facility level and decision-makers at the central level, as observed in Senegal.

- **Weak Accountability and Feedback Loops:** The motivation for civil servants to report high-quality data on time is significantly reduced if there is no accountability or consequence for poor performance. As documented in an assessment of M&E in Zimbabwe, a major demotivator was the lack of a feedback loop where ministries submitted performance reports, but "there was no feedback" on their performance.
- **Reporting Fatigue:** The proliferation of multiple, uncoordinated M&E systems, often driven by different donor requirements, can lead to reporting fatigue. This places an undue burden on staff, increasing the likelihood of errors, incomplete data, and late submissions.
- **Data Governance Deficiencies:** A lack of clear policies and procedures for data ownership, standardization, and quality control can lead to inconsistent data. The UN Women report on administrative data in Africa points to "discrepancies in the metadata structures" between different government departments as a key challenge to data integration and use.

7.3.4 Practical Strategies for Ensuring Data Quality and Timeliness

Overcoming these challenges requires a strategic and sustained effort that combines technical solutions with institutional reforms and a change in organizational culture.

- **Establish a Data Governance Framework:** A formalized data governance framework is essential. This framework, often led by the national M&E office, should define clear roles and responsibilities for data ownership, assign data stewards for different data sets, and establish standardized procedures for data collection, validation, and reporting. This ensures a consistent approach across the government and promotes accountability.

- **Invest in Continuous Training and Capacity Building:** Targeted and continuous training is critical. This should go beyond a single training session on the M&E system and include ongoing training on data collection methodologies, data entry protocols, and the use of data for decision-making. Training should be provided to all levels of the civil service, from the data collector in the field to the senior manager in the ministry.
- **Implement Robust Data Verification and Audits:** Regular data verification and quality audits are a non-negotiable best practice. These can include:
    - **Routine Data Quality Assessments (RDQAs):** As championed by organizations like MEASURE Evaluation, these are standardized, low-cost audits to routinely check the accuracy of data against its source documents at the facility or field level.
    - **Independent Audits:** In some cases, a third-party, independent audit of M&E data can be conducted to provide an unbiased assessment of its quality.
    - **Random Spot-Checks:** Regular, unannounced spot-checks by M&E officers can help to ensure that data is being collected correctly and on time.
- **Simplify and Standardize Reporting:** To combat reporting fatigue and improve data quality, governments should streamline their reporting requirements. This involves creating standardized reporting templates and timelines and, wherever

possible, integrating different reporting streams into a single, comprehensive M&E information system.

- **Create a Culture of Data Use and Accountability:** This is perhaps the most important long-term strategy. Leaders at all levels of government must be active consumers of M&E data. They must demonstrate that they use the data to inform decisions, provide feedback to reporting units, and hold their teams accountable for both results and the quality of the data they report. This creates a powerful incentive for civil servants to report high-quality, timely data, as they see that their efforts are valued and their work has a tangible impact. The assessment of M&E in Namibia highlighted that effective systems require "committed political and administrative leadership that demands and uses information."

A government's commitment to M&E is only as strong as its commitment to the quality and timeliness of its data. By addressing the challenges of human capacity, institutional weaknesses, and a lack of accountability, and by implementing a strategic mix of technical and cultural solutions, governments can build a robust data ecosystem that provides a reliable foundation for evidence-based decision-making and a more accountable public sector.

7.4.1 The Role of Technology and Innovation in Monitoring and Evaluation (M&E)

The landscape of Monitoring and Evaluation (M&E) has been fundamentally reshaped by the rapid advancements in technology and innovation. Once a slow, paper-based, and often disconnected process, M&E has evolved into a dynamic, data-driven, and real-time function. Technology now offers M&E practitioners the tools to collect, analyse, visualize, and report data with unprecedented speed, accuracy, and depth. This discussion explores the transformative role of technology in M&E, focusing on key innovations such as big data and geospatial

M&E, while also examining the benefits, challenges, and practical applications in various sectors.

7.4.2 The Transformative Role of Technology in M&E

The integration of new technologies marks a monumental shift from traditional M&E methods. The core purpose of M&E—to measure progress, assess outcomes, and inform decision-making—is now more efficient and effective thanks to digital tools.

From Paper to Pixels: The Digital Revolution in Data Collection

Historically, data collection relied on paper-based surveys, which were labour-intensive and prone to human error. The process of transcribing, cleaning, and analysing this data was time-consuming and often led to significant delays in reporting.

Today, mobile data collection (MDC) has become a cornerstone of modern M&E. Tools like Open Data Kit (ODK), KoBoToolbox, and SurveyCTO allow enumerators to use smartphones or tablets to collect data in the field. This approach offers several key advantages:

- **Real-time Data:** Data is uploaded to a central server as soon as an internet connection is established, providing stakeholders with a near-instantaneous view of project progress.
- **Improved Data Quality:** Digital forms can include built-in validation checks, such as dropdown menus and GPS coordinates, which minimize errors and inconsistencies.
- **Increased Efficiency:** The need for manual data entry is eliminated, saving time and resources. This is particularly valuable in large-scale projects, such as national censuses, as demonstrated by Kenya's 2019 census, which successfully utilized a mobile data collection tool to streamline the process.

Beyond mobile devices, the Internet of Things (IoT) is introducing "intelligent infrastructure," where sensors can be deployed to automatically collect data. For example, sensors on handwashing stations can track usage, or in-field monitors can measure water quality, providing continuous and passive data streams that were previously impossible to obtain.

7.4.3 The Power of Big Data in M&E

The proliferation of digital technologies has led to an explosion of data, often referred to as "big data." This encompasses not just the data collected through M&E activities but also vast datasets generated from other sources. When harnessed effectively, big data analytics can provide M&E professionals with a more holistic and nuanced understanding of project impact.

Defining and Utilizing Big Data

Big data is characterized by the "three Vs":

- **Volume:** The sheer scale of data generated from sources like social media, mobile phone records, and satellite imagery.
- **Velocity:** The speed at which this data is created and collected, often in real time.
- **Variety:** The diverse types of data, including structured data (e.g., survey results) and unstructured data (e.g., text from social media posts, videos, or photos).

In M&E, big data offers a powerful new lens for evaluation. For instance, analysing mobile phone call data can help track population movements after a natural disaster, allowing aid organizations to better target their interventions. Similarly, analysing public sentiment on social media platforms can provide insights into public opinion on a government policy or program.

Benefits of Big Data for M&E

- **Deeper Insights:** Big data allows for the analysis of patterns and trends that might be missed with traditional, smaller-scale datasets. This can help identify the underlying causes of program success or failure.
- **Predictive Analysis:** By analysing historical and real-time data, M&E professionals can build predictive models to forecast future outcomes, allowing for more proactive and adaptive management.
- **Cost-Effectiveness:** Using passively generated data, such as public transaction records or satellite imagery, can reduce the need for costly and time-consuming field surveys.

However, the use of big data in M&E is not without challenges, including concerns over data privacy, security, and the need for specialized skills to manage and analyze such large datasets.

7.4.4 Geospatial M&E: Visualizing Impact and Identifying Disparities

One of the most impactful technological innovations in M&E is the use of geospatial technologies, primarily Geographic Information Systems (GIS) and remote sensing. This approach integrates M&E data with geographical information, allowing practitioners to visualize project performance and outcomes on a map.

The Essentials of Geospatial M&E

Geospatial M&E is about answering the "where" questions of evaluation. By linking project data to specific locations, practitioners can:

- **Visualize Coverage and Reach:** Maps can clearly show which areas are being served by a project and which are not, helping to identify service gaps and target underserved populations.
- **Analyse Spatial Relationships:** Overlaying project data with other spatial datasets, such as demographics, infrastructure,

or environmental factors, can reveal crucial relationships. For example, a map might show that an education program's impact is lower in areas with poor road infrastructure.

- **Enhance Real-Time Monitoring:** Remote sensing technologies, such as satellite imagery and drone footage, allow for the large-scale and real-time monitoring of changes on the ground. This is particularly useful in sectors like agriculture (monitoring crop health) or disaster response (assessing damage to infrastructure).

For example, a health program can use GIS to map the locations of health clinics and the residences of beneficiaries to ensure equitable access. Similarly, in a natural resource management project, satellite imagery can be used to track deforestation over time, providing objective evidence of the project's impact on the environment.

7.4.5 Applications in the Public Sector and Development

Geospatial M&E is a powerful tool for governments and international development organizations. It helps in:

- **Resource Allocation:** By identifying areas with the greatest need, policymakers can allocate resources more strategically and efficiently.
- **Risk Management:** Maps can be used to identify areas prone to natural disasters or other risks, enabling proactive mitigation strategies.
- **Improved Communication:** Visualizations like interactive maps and dashboards make complex M&E findings more accessible and compelling for stakeholders, donors, and the public.

7.4.6 Challenges, Case Studies, and Future Outlook

While technology offers immense potential for M&E, its implementation is not without significant challenges, particularly in developing countries.

Key Challenges to Technological Adoption in M&E

- **Resource Constraints:** The initial cost of technology, including hardware, software, and training, can be prohibitive for many organizations with limited budgets.
- **Lack of Capacity:** M&E teams often lack the technical skills required to effectively use, manage, and analyse data from new technologies. This highlights the need for ongoing capacity building and training.
- **Data Quality and Integrity:** While technology can improve data quality, it also introduces new risks. Poorly designed digital tools can still produce "bad data," and the sheer volume of data can be overwhelming to manage and verify.
- **Ethical Considerations and Data Privacy:** The collection and use of vast datasets, especially those with personal information, raise serious ethical and privacy concerns. It is crucial to have robust data protection protocols in place, especially when working with vulnerable populations.

7.4.7 Case Study: Mobile Data Collection in African Agriculture

A practical example of technology's role is in agricultural development programs in Africa. Organizations have used mobile data collection tools to streamline the monitoring of farmer training and input distribution. By replacing paper forms with tablets, field officers can collect data on farmer participation, crop yields, and challenges faced. The real-time data allows program managers to quickly identify issues, such as a low adoption rate of a new farming technique in a specific region, and adjust their strategies accordingly.

7.4.8 The Future of M&E is Digital and Integrated

The role of technology in M&E is no longer a matter of choice but a necessity for organizations seeking to achieve greater effectiveness, efficiency, and accountability. Innovations like big data and geospatial M&E are not just tools; they are foundational elements that are reshaping the practice of evaluation itself. While challenges related to resources, capacity, and ethics remain, the benefits of a digitally integrated M&E system are undeniable. By embracing these technologies and addressing their associated challenges, organizations can move toward an M&E approach that is more adaptive, evidence-based, and ultimately, more impactful.

Chapter 8

The Political and Cultural Dimensions of Institutionalization of M&E

Introduction

The journey to institutionalizing a robust M&E system within government is a multifaceted endeavour that requires more than just technical expertise and organizational design. While a strong legal framework and a skilled workforce are essential, they are ultimately rendered ineffective without the bedrock of political will. The institutionalization of M&E is, at its core, a political process. It involves challenging established power dynamics, demanding accountability, and making difficult decisions about resource allocation based on evidence. In this complex landscape, the role of political leadership and "champions" of M&E is not just important; it is the single most critical factor for success. This chapter provides a comprehensive discussion on why political leadership is an indispensable prerequisite for M&E institutionalization, what constitutes a political champion, and how their actions shape the trajectory of a national M&E system, with a focus on real-world examples from the African context.

8.1.1 The Imperative of Political Leadership in M&E

M&E is inherently a political process because it concerns the management of public resources and the delivery of public services. It involves asking difficult questions and shining a light on areas of both success and failure, which can be uncomfortable for political leaders and public officials. For an M&E system to move from a theoretical concept to a functional reality, it requires a high-level political mandate to overcome the inherent resistance it will face.

- **Overcoming Institutional Inertia:** The public sector is often resistant to change, and M&E can be seen as a threat to established routines and power structures. A strong political champion uses their authority to overcome this inertia and provide a clear, unequivocal mandate that M&E is a government-wide priority.

- **Creating a Demand for Evidence:** In many public sectors, decision-making is based on political expediency or personal relationships rather than on evidence. A political champion changes this dynamic by publicly demanding and using M&E findings in their own decision-making, thereby creating a new institutional norm and a demand for evidence throughout the bureaucracy.
- **Providing the Necessary Authority:** For M&E findings to have teeth, the M&E body must have the authority to hold other ministries and departments accountable. This authority can only be granted and protected by a high-level political figure, such as a president or a prime minister.

8.1.2 What Constitutes a Political Champion of M&E?

A political champion of M&E is more than just a figurehead; they are an active advocate and protector of the M&E function. Their actions and commitment are what separate a successful M&E system from a failed one. A true champion of M&E can be identified by the following key characteristics and actions:

- **Providing a High-Level Mandate:** The most powerful signal a political champion can send is the creation of a central M&E institution within a high-level government office. The establishment of a Department of Performance Monitoring and Evaluation (DPME) in the Presidency of South Africa is a prime example of this. Similarly, the Office of the Prime Minister (OPM) in Uganda, taking the lead on the national M&E system, sent a clear message that M&E was a top-level priority, not just a technical or administrative function.
- **Leading by Example:** A champion leads from the front. They publicly demand to see M&E data on key national programs, discuss M&E findings in cabinet meetings, and use evidence

to inform policy and budget decisions. In South Africa, the Presidency's "delivery agreement" framework, which holds ministers accountable for a set of defined outcomes, is a clear example of leadership using M&E as a tool for accountability.

- **Resourcing the System:** A champion ensures that the M&E system is not a paper tiger. They allocate the necessary budget for M&E activities, including funding for a dedicated staff, training, and a robust information system. In Uganda, the government's commitment to M&E was institutionalized through a dedicated M&E budget line, a crucial step in ensuring the system's long-term sustainability.
- **Protecting the M&E Function:** An M&E system is only as credible as its independence. A champion protects the M&E body from political pressure and ensures that it is free to report both successes and failures without fear of retribution. This protection is vital for building a culture where staff feel safe to report honest data, even if it is unflattering.

8.1.3 The Tangible Impact of Leadership on M&E Institutionalization

The presence of a strong political champion has a tangible and transformative impact on the institutionalization of M&E.

- **Enhanced Accountability:** When a political champion makes M&E findings a key part of performance appraisals and public reporting, it fundamentally changes the game. Ministries and departments are incentivized to perform and report accurately, as their performance is now visible and consequential. This creates a powerful feedback loop that drives better service delivery and governance.
- **Improved Decision-Making:** A champion's commitment to evidence-based decision-making leads to a more rational and

effective allocation of public resources. Instead of relying on anecdotal information, policymakers can use credible M&E data to identify what works and what does not, leading to better policy design and implementation.

- **Building a Culture of Evidence:** The ultimate goal of a political champion is to create a culture of evidence where the demand for M&E information becomes a self-sustaining part of the public sector. When the bureaucracy sees that M&E is valued and used by leadership, the incentive to collect, manage, and use M&E data becomes widespread and enduring.

8.1.4 Case Studies of Political Leadership in the African Context

The African continent offers compelling case studies of the critical role of political leadership in M&E institutionalization.

- **South Africa:** The institutionalization of M&E in South Africa is arguably one of the continent's most successful examples. This success is directly tied to a high-level political mandate. The creation of the Department of Performance Monitoring and Evaluation (DPME) in the Presidency under former President Jacob Zuma was a landmark decision that signalled M&E as a core function of the executive. The DPME's mandate was to monitor the performance of all government departments against a set of national "service delivery agreements." This top-down, results-oriented approach, championed by the highest office, fundamentally changed the way the South African government operated. The fact that the DPME has survived multiple presidential transitions and continues to hold departments accountable is a testament to the strong political foundation that was built.

- **Uganda:** The institutionalization of M&E in Uganda is a powerful example of how political leadership can leverage partnerships to build a system and then take ownership of it. The M&E function, initially supported by the World Bank, was strategically championed by the government, leading to the creation of a dedicated M&E function within the Office of the Prime Minister (OPM). This high-level location provided the M&E body with the authority to oversee all government ministries. The government's commitment to M&E was further cemented by the allocation of a dedicated budget, a crucial action that moved the system from a donor-driven initiative to a fully owned government function.

8.1.5 The Complementary Role of Civil Society and International Partners

While a political champion is indispensable, they do not work in isolation. A successful M&E system requires a broader coalition of support from external actors.

- **Civil Society Organizations (CSOs):** CSOs, including think tanks and advocacy groups, play a crucial complementary role by serving as a check on government performance. By analysing and publicizing M&E findings, CSOs can hold political leaders accountable, thereby creating an external demand for quality data and transparency.
- **International Partners:** International development partners often play a critical role in the initial phases of M&E institutionalization, providing the technical expertise, funding, and capacity-building support needed to get a system off the ground. Their support, however, is most effective when it is aligned with a clear and sustained political mandate from

the host government.

8.1.6 Challenges and Recommendations for Sustaining Political Will

The primary challenge to M&E institutionalization is its sustainability. A system that is built on the will of a single champion is vulnerable to being dismantled or weakened when that champion leaves office. To ensure a system outlives a single administration, governments and their partners should:

- **Embed M&E in Legal and Policy Frameworks:** The ultimate form of political commitment is to embed M&E in legal frameworks, such as a national M&E Act or a clear M&E policy, as has been done in Zambia and Ghana. This makes M&E a legal requirement, not a political choice, making it much harder for future administrations to dismantle.
- **Build a Broad Coalition of Support:** Building a coalition of support from key institutions, such as the Ministry of Finance, Parliament, and independent audit bodies, makes the M&E system more resilient. When M&E is a requirement for budget allocation and legislative oversight, its importance is institutionalized across the government.
- **Institutionalize the Demand for Evidence:** The most sustainable M&E system is one where the demand for evidence is not just top-down but also bottom-up. The goal is to create a culture where civil servants and citizens alike see the value in using evidence to drive results, making M&E an indispensable part of good governance.

The institutionalization of M&E is a political process that requires a sustained and active political champion. By providing a clear mandate, leading by example, and building a broad coalition of support, a political champion can transform M&E from a technical exercise into a powerful tool for accountability, improved decision-making, and better governance for their citizens.

8.2 Addressing the "Culture of Resistance" to M&E in Public Institutions

The institutionalization of M&E is a journey that is often beset by a silent but formidable barrier: a "culture of resistance." While high-level political champions can provide the top-down mandate for a national M&E system, they cannot, on their own, change the deeply ingrained attitudes and behaviours of civil servants who must operate it. This section delves into the reasons behind this resistance and provides a set of practical, actionable strategies for shifting the institutional culture from one of fear and opposition to one of ownership and a genuine demand for evidence.

8.2.1 The "Culture of Resistance": A Diagnostic View

Before a culture of resistance can be addressed, its root causes must be understood. The opposition to M&E is not arbitrary; it is often a rational response to how M&E is perceived and implemented within the public sector. The main reasons for resistance can be categorized as follows:

- **Fear of Accountability and Failure:** For many civil servants, M&E is perceived as a punitive tool designed to find fault, assign blame, and expose failures. This perception is often rooted in a culture where performance reviews are seen as a source of punishment rather than an opportunity for learning and growth. As a result, staff may be incentivized to underreport poor performance or even manipulate data to avoid negative consequences. The M&E function is seen as a "threat to power and authority" rather than a tool for improvement.

- **Perception as a Bureaucratic Burden:** Civil servants are often overworked and under-resourced. In this context, M&E can be seen as yet another administrative burden, a "parallel, donor-driven activity" that adds to their workload without providing any tangible benefits. The reporting requirements are often not aligned with their daily work, and the data they collect is perceived as disappearing into a black hole, never to be used for decision-making. This lack of a feedback loop is a major demotivator for timely and accurate reporting.
- **Lack of Ownership and Understanding:** Resistance often stems from a lack of understanding of what M&E is and a feeling that the system was "imposed from the top." When staff are not consulted during the design of an M&E system, they are less likely to take ownership of it. Furthermore, a lack of targeted training can leave civil servants feeling incompetent and ill-equipped to perform M&E tasks, leading to a natural resistance to a function they do not understand.

8.2.2 Shifting the Culture: From Resistance to Ownership

Overcoming a culture of resistance requires a strategic and deliberate approach that re-frames M&E from a coercive compliance function to a valued learning tool. The following strategies are crucial for this cultural shift:

- **Re-frame M&E as a Learning Tool, Not a Punitive One:** The first step is to change the narrative around M&E. Political champions and M&E leaders must publicly and consistently frame M&E as a powerful source of "organizational learning and continuous improvement." For instance, an M&E report should not just highlight where a program failed but should also provide recommendations for how it can be improved. A successful initiative in **South Africa** involved reframing M&E from a "culture of compliance to a culture of performance," where data was used to identify what worked and what didn't, rather than just to assign blame.
- **Demonstrate the Value of M&E to Civil Servants:** The most effective way to overcome resistance is to demonstrate how M&E can help civil servants do their jobs better. This can involve providing them with real-time dashboards that show their progress against targets, providing evidence that can be used to justify resource requests, or using M&E findings to improve program design. When civil servants see that M&E can lead to better outcomes and a more efficient workflow, they are more likely to embrace it.
- **Build Capacity and Provide Continuous Support:** Resistance often comes from a place of fear and a lack of competence. This can be addressed through continuous, targeted training that is "aligned with the specific roles and responsibilities" of civil servants. This training should be practical and hands-on, building confidence in their ability to perform M&E tasks. Furthermore, a strong support system, such as a help desk or a community of practice, can provide a

safety net for civil servants as they navigate a new system.

- **Foster a Participatory Approach:** M&E systems that are designed without input from the people who will operate them are almost certain to fail. A best practice is to involve staff from all levels of the government in the design and planning of the M&E system. This participatory approach, as advocated by **UNICEF**, helps to build a sense of ownership and commitment from the ground up, making staff feel that the system is "theirs" rather than a tool imposed on them.
- **Link M&E Findings to Tangible Incentives:** People respond to incentives. While a culture of learning is the ultimate goal, tangible incentives can be a powerful tool for overcoming initial resistance. These can be both financial and non-financial, such as linking M&E performance to promotions, providing public recognition for good M&E practice, or using M&E findings to allocate performance bonuses. The use of M&E to inform budget decisions in South Africa is a powerful example of an incentive that helped to change the institutional culture.

8.2.3 Case Studies of Cultural Change in African Public Institutions

Several African countries provide compelling examples of how a culture of resistance can be overcome through strategic interventions.

- **Ghana:** A study on M&E in Ghana's public sector highlighted the challenges of resistance and proposed a "change management approach." This approach included a broad-based strategy for training and engaging all stakeholders, from senior management to frontline staff. The study emphasized the importance of fostering a dialogue around M&E to build understanding and ownership.

- **Uganda:** In Uganda's health sector, a key challenge was getting health workers to embrace a new, technology-driven data system. This was addressed through a participatory approach where health workers were involved in the design and piloting of the new system. This helped to build trust and a sense of ownership, as they felt that their input was valued.
- **South Africa:** The journey of institutionalizing M&E in South Africa provides a powerful case study in cultural transformation. While the political champions at the top provided the mandate, the real work involved getting the civil service on board. The use of performance information to inform budget decisions was a major incentive. The government also used a "common language" and standardized frameworks to make M&E accessible and understandable to all civil servants, helping to reduce the perception of it as an arcane, top-down function.

In summary, addressing a culture of resistance is not a one-off event but a sustained, strategic effort that is just as important as building the M&E system itself. By re-framing M&E from a punitive tool to a learning tool, by demonstrating its value to civil servants, by fostering a participatory approach, and by providing tangible incentives, governments can transform the institutional culture and build a public service that is not only capable of doing M&E but is also genuinely committed to using evidence to drive better results.

8.3 The Link Between M&E and Accountability: How Findings Are Used (or Not Used)

The ultimate purpose of a national M&E system is not merely to produce reports; it is to create a more accountable and effective government. The M&E function provides the evidence—the data, findings, and recommendations—that can be used to hold public officials accountable for their performance and to inform better policy decisions. The link between M&E and accountability, however, is not automatic. The space between producing an M&E report and acting on its findings is what is often referred to as the "accountability gap." This section provides a comprehensive discussion on the vital link between M&E and accountability, the reasons why this link often fails, and the practical strategies that governments can use to ensure M&E findings are not just seen, but used.

8.3.1 The M&E-Accountability Nexus

Accountability in the public sector is the obligation of an individual or an organization to justify their actions to a higher authority. M&E provides the necessary information to enable this process. The link between M&E and accountability can be viewed in two dimensions:

- **Horizontal Accountability:** This is the internal accountability within the government. M&E findings enable a central authority, such as the President's Office or the National Treasury, to hold line ministries and departments accountable for their performance against national development goals and allocated budgets. For example, a central M&E unit can use performance data to question a ministry on why it is not meeting its targets for school enrolment.

- **Vertical Accountability:** This is the external accountability of the government to its citizens. M&E findings, when made public, provide citizens, civil society organizations, and Parliament with the information they need to demand better performance from their government. For example, a publicly released evaluation report on a social protection program can empower citizens to hold their elected representatives accountable for its effectiveness.

8.3.2 The Challenge of "M&E Findings Not Used": The Accountability Gap

Despite the clear potential, a common challenge in many public sectors is the non-use of M&E findings. This "accountability gap" arises from a combination of political, institutional, and technical factors. The main reasons why evaluation findings often go unutilized include:

- **Lack of a Political Champion:** As discussed in the previous chapter, without a high-level political champion, M&E findings are easily ignored. If a president or a minister does not demand to see and use M&E findings, there is little incentive for civil servants to act on them.
- **Poorly Communicated Findings:** M&E findings are often buried in long, technical reports that are inaccessible to policymakers and the public. These reports are often written in academic language, lack clear recommendations for action, and are not designed for a policy-making audience. The challenge is not a lack of data, but a failure to communicate it effectively.
- **Absence of a Formal Follow-up Mechanism:** In many public sectors, there is no formal system for departments to respond

to M&E recommendations. An evaluation report may be produced, but if there is no institutionalized mechanism to track the implementation of its recommendations, it is often forgotten and left to gather dust on a shelf.

- **Mismatch of Timing:** M&E findings often arrive too late to inform the budget and planning cycles. A major evaluation report released in the middle of a financial year, after the budget has already been approved, may have excellent recommendations, but they cannot be acted upon until the next budget cycle, by which time they may no longer be relevant.

8.3.3 Strategies for Closing the Accountability Gap

Closing the accountability gap requires a strategic and sustained effort that institutionalizes the use of M&E findings. The following strategies are crucial for ensuring that M&E is a tool for action, not just an academic exercise.

- **Communicate for Action, Not Just for Information:** The format and content of M&E reports must be designed for their intended audience. Best practices include:
    - **Executive Summaries and Policy Briefs:** Producing short, two-to-three-page summaries that highlight key findings and clear recommendations for action.
    - **Data Visualization:** Using dashboards, infographics, and interactive maps to make complex data accessible and compelling for policymakers and the public.
    - **Dissemination Events:** Holding workshops and briefings with policymakers to present findings and facilitate a dialogue around their implications.
- **Institutionalize Follow-up Mechanisms:** The most effective

way to ensure that M&E findings are used is to create a formal system to track the implementation of recommendations. A "management response system" is a powerful tool for this. This system requires government departments to formally respond to evaluation recommendations, outlining which recommendations they accept and what actions they will take to implement them. The central M&E body then monitors the implementation of these actions, thereby creating a clear paper trail for accountability.

- **Link M&E Findings to the Budget and Planning Cycle:** The most powerful incentive for government departments to use M&E findings is to link them to the budget. This is a core function of horizontal accountability. A central M&E body can work with the Ministry of Finance to use performance data and evaluation findings to inform budget allocations. For example, a ministry that consistently fails to meet its targets may have its budget reduced or reallocated to a better-performing program.
- **Empower a Central Authority to Enforce Accountability:** A central M&E body, ideally located in a powerful office like the Presidency or the National Treasury, is crucial for enforcing the use of M&E findings. This central body provides the authority and oversight needed to hold departments accountable and to ensure that M&E is not seen as an optional extra but a core function of governance.

8.3.4 Case Studies of M&E and Accountability in the African Public Sector

Several African countries provide compelling case studies of both the challenges and successes of linking M&E to accountability.

- **South Africa:** South Africa provides one of the most powerful examples of horizontal accountability. The Department of Performance Monitoring and Evaluation (DPME), located in the Presidency, conducts rigorous evaluations and performance reviews. These findings are then submitted to the National Treasury, which uses them to inform the budget allocation process. A ministry that consistently performs poorly or fails to respond to M&E recommendations may have its budget reduced. This direct link between performance and funding provides a powerful, institutionalized incentive for accountability.
- **Kenya:** Kenya's **National Integrated Monitoring and Evaluation System (NIMES)** provides a good example of a government attempting to promote vertical accountability. The NIMES system produces regular reports on the progress of national development goals, and these reports are made accessible to the public. This empowers citizens and civil society to use this data to advocate for better performance and to hold their elected representatives accountable.
- **Uganda:** In Uganda, the **Office of the Prime Minister (OPM)**, which is the custodian of the national M&E system, has used its authority to ensure that M&E findings are used for decision-making. The OPM's M&E function produces annual reports on government performance, and these reports are discussed in Cabinet meetings. This top-down, political use of M&E findings creates a strong demand for accountability from all government ministries.

The institutionalization of M&E is not complete until its findings are routinely used for accountability and evidence-based decision-making. By strategically addressing the accountability gap through effective communication, institutionalized follow-up mechanisms, and a direct link to the budget, governments can ensure that M&E is a powerful tool for improving performance, building public trust, and delivering on their promises to their citizens.

8.4 Promoting a Learning Culture Rather Than a Blame Culture

The institutionalization of M&E within government is a journey of systemic change that transcends the creation of legal frameworks, organizational structures, and technical systems. The true measure of a successful M&E system lies in its ability to transform the institutional culture itself. This final section explores the most critical cultural shift required for M&E to achieve its full potential: moving from a "blame culture" to a "learning culture." A blame culture sees M&E as a punitive tool to find fault, while a learning culture views it as an indispensable tool for continuous improvement and adaptive management. This discussion will delve into the negative consequences of a blame culture, provide a set of actionable strategies for fostering a learning-oriented environment, and highlight real-world examples of this cultural shift in the African public sector.

8.4.1 The Great Cultural Shift: From Blame to Learning

The culture in which an M&E system operates is a critical determinant of its success. It defines how data is perceived, how findings are used, and how public officials engage with the entire process.

- **The Blame Culture:** In a blame culture, the primary focus is on avoiding failure and assigning responsibility when things go wrong. M&E is viewed as a threat—a mechanism for "blame and punishment." In this environment, M&E data is often manipulated to reflect success, and failures are hidden to avoid negative consequences. Risk-aversion is the norm, and innovation is stifled for fear of making a mistake.

- **The Learning Culture:** In contrast, a learning culture views M&E as a powerful source of knowledge and a tool for continuous improvement. Failure is not seen as an endpoint but as a valuable opportunity to learn and adapt. M&E findings are discussed openly and constructively, with the goal of understanding "why" a particular outcome occurred and "what can be done differently" in the future. This culture fosters a sense of ownership, trust, and a genuine commitment to using evidence to achieve better results.

8.4.2 The Negative Consequences of a Blame Culture

A blame culture has profound and tangible negative impacts on the public sector's ability to perform and learn.

- **Poor Data Quality and Integrity:** When civil servants fear punishment, they are incentivized to provide data that presents a positive picture, even if it is inaccurate. This manipulation of data renders the M&E system useless, as it no longer reflects the reality on the ground.
- **Stifled Innovation and Adaptive Management:** A blame culture makes managers and staff risk-averse and hesitant to try new approaches or deviate from a pre-defined plan. This stifles innovation and prevents organizations from making the necessary "mid-course corrections" that are often required for complex development programs.

- **Loss of Institutional Memory:** When failures are hidden, the lessons that could have been learned from them are also hidden. This leads to a situation where the same mistakes are repeated over and over again, as the institution loses the ability to learn from its past.
- **Increased Resistance to M&E:** The perception of M&E as a punitive tool is a primary driver of the "culture of resistance" discussed in the previous chapter. This resistance, in turn, undermines the entire M&E institutionalization effort.

8.4.3 Strategies for Fostering a Learning Culture

Fostering a learning culture is a long-term, strategic effort that requires a combination of top-down leadership and bottom-up engagement. The following strategies are crucial for making this cultural shift:

- **Lead by Example and Publicly Acknowledge Failure:** The most powerful signal that a government can send is for its leaders to publicly acknowledge and discuss both successes and failures as learning opportunities. When a President or a Minister speaks openly about a program that did not work as intended and the lessons learned from it, it creates a "safe space" for civil servants to do the same. This top-down approach is essential for demonstrating that M&E is genuinely valued for learning, not just for accountability.
- **Create Safe Spaces for Dialogue and Reflection:** For a learning culture to take root, organizations must institutionalize forums where staff can openly and honestly discuss challenges without fear of punishment. This can be done through:
    - **"Lessons Learned" Workshops:** These workshops, held at the

end of a project, bring all stakeholders together to reflect on what worked and what did not.

- **After-Action Reviews (AARs):** These are informal, peer-to-peer discussions held immediately after a major activity to capture lessons while they are still fresh.
- **Communities of Practice:** These forums allow M&E practitioners from across the government to share experiences, discuss common challenges, and learn from each other.

- **Link M&E to Adaptive Management:** The M&E system must be designed to be useful for adaptive management. M&E findings should be used to make real-time adjustments to programs and policies, rather than just being presented in a final report. When civil servants see that the data they collect is being used to make "mid-course corrections" that lead to better results, they are more likely to see the value of M&E as a tool for improvement.
- **Reward Learning, Not Just Success:** The incentive structure must be changed to reward learning and innovation, not just success. This can be done by:
    - **Recognizing the Identification of Problems:** Providing recognition to a team that used M&E to identify a fundamental problem in a program, even if the program was not a success.
    - **Linking M&E to Career Development:** M&E skills and a demonstrated ability to use evidence for learning should be key criteria for promotions and career advancement.
- **Communicate M&E Findings for Learning:** M&E reports and communications must be reframed to focus on "why" a particular outcome occurred. Reports should be accessible, policy-relevant, and provide actionable recommendations for the future. The focus should be on generating knowledge and insights, not just on presenting performance numbers.

8.4.4 Case Studies from the African Public Sector

Several African countries and institutions have made deliberate efforts to promote a learning culture within their M&E systems, providing valuable case studies.

- **South Africa:** The journey of M&E in South Africa is a great example of a country trying to move beyond a blame culture. While the initial focus of the DPME was on accountability and performance numbers, there has been a growing emphasis on using M&E for learning. The government has created forums for departments to discuss and learn from evaluation findings, thereby fostering a more open and constructive dialogue around performance.
- **Ghana:** A study on M&E institutionalization in Ghana highlighted the need for a "change management approach" to overcome resistance. The study recommended a broad-based strategy for training and engaging all levels of the civil service to build a shared understanding of M&E as a tool for learning and improvement. This points to the need for a sustained effort to change mindsets from the ground up.
- **International Organizations:** The use of adaptive management by many international organizations and their partners in Africa provides a powerful example of a learning culture in practice. In these programs, M&E data is used to make real-time adjustments to projects, and regular "lessons learned" workshops are a standard part of the project cycle. This demonstrates that a learning culture is not a theoretical concept but a practical reality.

8.4.5 The End Goal of Institutionalization

The ultimate goal of M&E institutionalization is not the existence of a national M&E policy or a department in the Presidency. It is the creation of a public sector culture that is self-correcting, adaptive, and genuinely committed to using evidence for continuous improvement. By making the shift from a blame culture to a learning culture, governments can transform M&E from a bureaucratic exercise into a dynamic force for better governance, innovation, and ultimately, better results for their citizens. This cultural shift is the final frontier of M&E institutionalization and the key to building a truly results-oriented public sector.

Chapter 9

The Role of External Actors

Introduction

In today's development and humanitarian landscape, Monitoring and Evaluation (M&E) have become critical tools for ensuring accountability, learning, and evidence-based decision-making. While internal actors such as program staff and implementing agencies play a central role in designing and executing M&E systems, external actors bring unique value that strengthens the integrity, objectivity, and effectiveness of these processes. External actors—including consultants, independent evaluators, academic institutions, donor agencies, and civil society organizations—serve as impartial observers, technical experts, and accountability agents. Their involvement is especially crucial in contexts where local capacity is limited, political interests may compromise internal assessments, or complex multi-stakeholder programs demand an independent eye.

The presence of external actors can enhance credibility by reducing potential biases in data collection and reporting. They also contribute technical skills, global perspectives, and comparative insights that help to contextualize findings within broader development frameworks. Moreover, donors often require independent evaluations as part of funding agreements, reinforcing the demand for rigorous, transparent assessments. However, the engagement of external actors is not without challenges. Issues such as limited contextual understanding, high costs, and possible misalignment with local priorities can hinder the effectiveness of externally driven M&E efforts. Therefore, their role must be balanced with participatory approaches that respect local knowledge, promote capacity building, and foster collaboration with internal stakeholders.

This chapter explores the multifaceted role of external actors in M&E, highlighting their contributions, limitations, and the principles that guide their engagement for ethical and impactful evaluation practice.

9.1 The Influence of International Development Partners

The institutionalization of Monitoring and Evaluation (M&E) systems in many African countries is a complex process shaped not only by internal political will and institutional dynamics but also by the powerful influence of external actors. International development partners, including multilateral organizations like the World Bank and the UN, bilateral donors, and international NGOs, play a multifaceted role in this process. While their involvement can be a catalyst for building robust M&E systems and fostering a culture of accountability, it also presents significant challenges. This section provides a comprehensive discussion on the influence of international development partners on national M&E systems, examining their varied roles, the positive and negative impacts of their engagement, and offering recommendations for building a more collaborative and effective partnership.

9.1.1 The Varied Roles of International Development Partners

International development partners influence national M&E systems in a variety of ways, serving as more than just a source of funding. Their influence can be both direct and indirect, shaping the very nature of M&E within a country.

- **As Providers of Technical and Financial Support:** This is arguably the most common role. Partners provide catalytic funding to initiate M&E initiatives, develop national M&E policies, and purchase the necessary software and hardware. They also offer technical assistance, bringing in experts to provide training, build systems, and assist with complex data analysis.
- **As Drivers of M&E Agendas:** Many development partners operate under a results-based management framework, which

requires their funded projects to demonstrate measurable progress and impact. This often compels governments to adopt certain M&E standards and practices to meet donor reporting requirements. International frameworks like the **Sustainable Development Goals (SDGs)** also influence national M&E systems by encouraging countries to adopt a common set of indicators for global reporting.

- **As Users and Disseminators of M&E Findings:** Partners are a critical audience for M&E data. They use a government's M&E findings to inform their own programming, evaluate the effectiveness of their aid, and advocate for policy changes. They can also play a role in disseminating M&E findings to a wider global audience, including through their own reports and publications.

9.1.2 The Positive Influence of International Development Partners

When engaged constructively and in alignment with national priorities, international development partners can be a powerful force for strengthening national M&E systems.

- **Catalytic Funding:** External funding can provide the initial push for M&E initiatives that governments, due to other competing priorities, may not have been able to fund on their own. The establishment of Uganda's national M&E system within the Office of the Prime Minister (OPM), for instance, was initiated with significant technical and financial support from the World Bank and other partners. This support was instrumental in getting the system off the ground.
- **Driving Standards and Best Practices:** Development partners can introduce international best practices and methodologies for M&E, helping to professionalize the field

within government. They can provide access to high-quality training materials, state-of-the-art software, and expertise in areas such as impact evaluation. This helps to build a more rigorous and credible M&E system.

- **Capacity Building:** Partners play a crucial role in building the skills of national M&E practitioners. The World Bank's support for M&E in Rwanda, for example, has included hands-on technical assistance in areas such as data analysis and report writing. This type of targeted, sustained capacity building helps to create a pool of skilled government officials who can manage and lead their own M&E systems.
- **Strengthening Accountability:** By consistently demanding evidence of results, partners can act as an external check on government performance. This creates a powerful incentive for governments to build credible M&E systems to demonstrate to both their own citizens and the international community that aid is being used effectively.

9.1.3 The Negative and Unintended Consequences of External Influence

While the potential for positive influence is great, the engagement of international partners can also lead to significant and unintended consequences that undermine, rather than strengthen, national M&E systems.

- **The "Parallel M&E Systems" Problem:** This is perhaps the most common and damaging unintended consequence. Individual donors often have their own specific, uncoordinated reporting requirements and indicators that do not align with a country's national M&E framework. This forces government ministries to create separate, parallel M&E systems to meet the demands of different partners. This problem, which was a key driver of the Paris Declaration on Aid Effectiveness, leads to a "heavy burden on government staff" and creates a fragmented and uncoordinated M&E landscape. Kenya's National Integrated Monitoring and Evaluation System (NIMES), for example, has struggled to coordinate with the numerous parallel reporting requirements of different donor partners.

- **M&E as a "Donor-Driven" Function:** When external partners provide too much financial and technical support, M&E can be perceived as a "donor-driven" function rather than a government priority. This undermines government ownership and can lead to a situation where M&E systems are seen as a requirement to access funding rather than a tool for improving governance and service delivery.
- **Focus on Donor Priorities, Not National Priorities:** Donors often come with their own set of programmatic priorities and indicators. This can lead to a situation where a government's M&E system is skewed toward collecting data that is relevant to a donor's agenda, even if it is not the most relevant for national development goals.
- **High Transaction Costs:** The burden of multiple, overlapping reporting requirements on government staff and the costs associated with managing different donor expectations can be immense. This can divert resources and attention away from building a single, coherent national M&E system.

9.1.4 Case Studies of Donor Influence in Africa

The experiences of African countries provide compelling examples of both the benefits and challenges of external influence on M&E.

- **Uganda: A Model of Successful Partnership:** The institutionalization of M&E in Uganda is a good example of how a successful partnership can work. With the support of the World Bank and other donors, the government established a national M&E system in the Office of the Prime Minister (OPM). While donors provided the initial push and resources, the government took full ownership of the system, embedding it within the national planning and budgeting process. This approach ensured that M&E was not a donor-driven function but a government-owned tool for accountability and decision-making.
- **South Africa: Government Ownership from the Start:** South Africa's experience is different. The institutionalization of M&E was driven by strong political will from the Presidency, with the creation of the Department of Performance Monitoring and Evaluation (DPME). In this case, the role of international partners was less about providing catalytic funding and more about providing "peer support," sharing knowledge, and advocating for best practices. This is an excellent example of a government-owned system where external partners supported, rather than drove, the M&E agenda.
- **Kenya and Ghana: The Challenges of Coordination:** The experiences of countries like Kenya and Ghana highlight the challenges of a fragmented donor landscape. In both countries, despite having national M&E policies and frameworks, the existence of numerous donor-specific reporting requirements has led to the proliferation of parallel M&E systems, making it difficult for the government to get a coherent, single view of national performance. This underscores the need for better coordination and

harmonization among donors themselves.

9.1.5 Recommendations for Effective Partnership

To ensure that the role of international development partners is constructive and strengthens, rather than undermines, national M&E systems, both governments and partners must adopt a new approach to collaboration.

- **Governments Must Lead and Own the Agenda:** Governments must take the lead in developing their national M&E policies and frameworks. They must be clear about their M&E priorities and be firm in requiring all partners to align with them.
- **Partners Must Align with National Systems:** Development partners must adhere to the principles of the **Paris Declaration on Aid Effectiveness**, which calls for aid harmonization and for partners to use a country's own M&E systems. They must resist the temptation to impose their own reporting requirements and instead work to strengthen the national system.
- **Shift from Short-Term Assistance to Long-Term Capacity Building:** Partners should move away from providing short-term, expert-driven technical assistance and instead invest in long-term capacity building that empowers government officials with the skills to design, manage, and use their own M&E systems.
- **Build a Culture of Shared Learning:** The relationship between a government and its partners should be one of shared learning. Both sides should openly discuss what is working and what is not, and use M&E findings to collectively improve the effectiveness of aid and national development programs.

The influence of international development partners on national M&E systems is a powerful but delicate force. When their engagement is aligned with national priorities, it can be a catalyst for strengthening M&E and improving governance. When it is not, it can create fragmented and unsustainable systems. The key to success lies in a collaborative partnership built on the principles of government ownership, mutual trust, and a shared commitment to building robust, long-term M&E capacity.

9.2 Donor Requirements: A Double-Edged Sword in Institutionalization

The influence of international development partners on national M&E systems is most tangibly felt through their specific reporting requirements. These requirements, which are often a prerequisite for funding, can be a powerful force that either catalyses or complicates the institutionalization process. While they are designed with the best intentions—to ensure transparency, accountability, and the effective use of resources—their implementation can create a complex web of challenges for governments in the Global South. This discussion will delve into the dual nature of donor requirements, examining how they can both help and hinder the institutionalization of robust, government-owned M&E systems.

9.2.1 How Donor Requirements Can Help the Institutionalization Process

In many contexts, donor requirements have served as a critical catalyst for governments to take M&E seriously. When used strategically, they can provide the necessary push to overcome internal inertia and build a foundation for a results-oriented public sector.

- **Driving a Results-Oriented Culture:** The international emphasis on results-based management, driven by donors, has been a key factor in the institutionalization of M&E in many African countries. Donors' demand for clear, measurable outcomes provides the political impetus for governments to shift from a focus on inputs and activities to a focus on results. In contexts where a strong national demand for M&E is initially absent, donor requirements can serve as a powerful external stimulus, thereby "stimulating the development of M&E practice."

- **Providing a Common Language and Standards:** While a multiplicity of donor requirements can be a burden, a shared set of M&E standards and indicators among a group of donors can help to standardize M&E practices within a country. By pushing for a common framework (e.g., in line with the Sustainable Development Goals), donors can help to professionalize the M&E function and create a common language for performance measurement.
- **Enhancing Accountability:** Donor requirements for performance reporting can strengthen both horizontal (within government) and vertical (to the public) accountability. By demanding regular reports on progress and outcomes, donors create a demand for credible performance data, which can be used by a central M&E unit to hold line ministries accountable. When this data is made public, it can also be used by citizens and civil society to hold the government to account, thereby closing the accountability loop.
- **Incentivizing Capacity Building:** The need to meet donor reporting requirements can act as a strong incentive for governments to invest in M&E capacity building. The need for a government to produce credible performance reports often leads to investments in staff training, the development of M&E frameworks, and the establishment of dedicated M&E units. In this way, donor requirements can trigger a virtuous

cycle of investment in M&E capacity.

9.2.2 How Donor Requirements Can Hinder the Institutionalization Process

Despite their potential to help, donor requirements can also be a significant impediment to the development of a coherent, government-owned M&E system. The challenges often arise from a lack of coordination among donors and a focus on compliance over genuine learning and improvement.

- **The "Parallel M&E Systems" Problem:** This is arguably the most damaging unintended consequence of donor requirements. When individual donors have their own specific, often uncoordinated, reporting requirements, governments are forced to create separate, parallel M&E systems to meet the demands of each partner. This leads to a fragmentation of the M&E landscape, with ministries and departments dedicating significant resources to tracking indicators and producing reports that are relevant to a donor but not necessarily to the national M&E framework. A rapid assessment of the M&E burden in the health sector in some African countries, for example, has shown that international reporting requirements can add a significant number of indicators to national M&E plans, leading to "fragmented data collection, uncoordinated efforts... and a heavy burden on government staff."
- **The "Tyranny of Indicators": Distorting National Priorities:** Donors often come to the table with a pre-defined set of indicators that are tied to their own programmatic agendas. When a government's M&E system becomes focused on reporting on these donor-driven indicators, it can lead to a neglect of broader national development priorities. A case

study on social protection in Zimbabwe, for example, found that the policy process was "almost entirely donor-driven," with the government acting as a "passive observer." This example shows how a system that is meant to serve national priorities can be distorted to serve external ones, leading to a loss of ownership.

- **Fostering a Culture of Compliance over Learning:** When M&E is seen primarily as a requirement to satisfy a donor, the institutional focus shifts from using findings for learning and improvement to simply checking a box for compliance. In this environment, M&E data may be manipulated to present a positive picture, and failures are hidden to avoid jeopardizing future funding. This undermines the quality and integrity of the data and, as one study noted, renders the M&E system of "little value for learning."
- **High Transaction Costs:** The burden of multiple, overlapping reporting requirements on government staff is immense. Each donor may require a different reporting format, a different set of indicators, and a different reporting frequency. This creates high transaction costs for a government that is already resource-constrained. The time and effort spent on satisfying donor requirements could be better spent on strengthening a single, coherent national M&E system and using its findings for effective decision-making.

9.2.3 Recommendations for an Effective Partnership

The influence of donor requirements is a delicate balance. When they are aligned with national priorities and coordinated among partners, they can be a powerful force for institutional strengthening. When they are not, they can be a source of fragmentation, a reporting burden, and a major impediment to building a sustainable, government-owned M&E system.

To ensure that donor requirements are a force for good, a new approach to collaboration is needed, one that is built on the principles of the **Paris Declaration on Aid Effectiveness.**

- **Government Leadership:** Governments must take the lead in developing their own M&E frameworks, defining national priorities, and requiring all donors to align with a single national system.
- **Donor Harmonization:** Donors must resist the temptation to impose their own reporting requirements and instead work together to harmonize their M&E needs and use a country's own M&E system.
- **Shift from Compliance to Learning:** Both governments and donors must foster a culture where M&E is valued not just for accountability and compliance, but also as a tool for learning, adaptation, and continuous improvement.

The institutionalization of M&E is not a passive process. It is a journey where the influence of donor requirements must be actively managed to ensure that they become a force for empowerment and ownership, not a source of fragmentation and dependency.

9.3 The Importance of Country-Led M&E Systems and Reducing Donor Dependency

While international development partners have been a powerful force in promoting M&E institutionalization, the long-term success and sustainability of these systems hinge on a single, paramount principle: country ownership. A monitoring and evaluation system that is genuinely "country-led" is one that is designed, funded, managed, and used by the government itself, in service of its own national development goals. The journey from a donor-dependent model to a country-led one is arguably the final and most critical step in the institutionalization process. This chapter explores the imperative of a country-led M&E system, the perils of donor dependency, and the strategic pathways for governments to build self-reliant and sustainable M&E functions.

9.3.1 The Imperative of a Country-Led M&E System

A country-led M&E system is not merely a preference; it is an imperative for effective and sustainable governance. Its importance is rooted in several key dimensions:

- **National Sovereignty and Alignment:** A country-led system ensures that M&E efforts are directly aligned with a nation's own development agenda, policies, and priorities. When the M&E framework is a product of domestic political and institutional processes, it is more likely to track indicators that are most relevant to the country's unique challenges and goals, rather than external agendas. This alignment is critical for evidence-based policy-making that is genuinely responsive to national needs.
- **Long-Term Sustainability:** An M&E system that is funded and managed by the government is inherently more sustainable. While donor funding can be a valuable catalyst, it is often short-term and project-based. When funding ends, a donor-dependent system can collapse, taking with it all the institutional memory, capacity, and infrastructure that was built. A country-led system, funded through domestic resources, is not vulnerable to these external shocks.
- **Legitimacy and Trust:** A government-led M&E system is more likely to be trusted and used by domestic policymakers, civil servants, and citizens. When the system is seen as an integral part of the governance structure, its findings carry greater weight and are more likely to be used for decision-

making. It also strengthens vertical accountability by providing citizens with a legitimate source of information on government performance, thereby building public trust.

9.3.2 The Perils of Donor Dependency

The opposite of a country-led system is one characterized by donor dependency, a state that poses significant threats to the institutionalization and sustainability of M&E. The challenges of donor dependency include:

- **Unsustainable Systems:** The "project mentality" that often accompanies donor funding is a major threat to sustainability. In this scenario, government staff and policymakers may see M&E as a short-term, donor-driven project rather than a core, permanent function of governance. Once donor support is exhausted, the system and the capacity built around it are at high risk of withering away.
- **The Fragmentation of M&E:** As discussed in the previous chapter, a lack of a single, country-led M&E framework often leads to the proliferation of "parallel M&E systems." When each donor has its own reporting requirements, government ministries are forced to dedicate resources to meeting multiple, uncoordinated demands. This fragmentation wastes resources, creates a heavy reporting burden, and prevents the government from having a single, coherent view of its own performance.
- **A Culture of Compliance, Not Action:** A donor-dependent system can foster a culture of compliance where the primary objective of M&E is to satisfy external reporting requirements. In this environment, the focus is on "checking the box" rather than on using M&E findings for learning, adapting programs, and making evidence-based decisions.

This undermines the very purpose of M&E and prevents it from being a dynamic force for change.

- **Distorted National Priorities:** A key peril of donor dependency is the risk that a country's M&E agenda will be distorted to serve donor priorities over national ones. A study on the health sector in Ghana, for instance, found that donor financing historically "targeted certain programmatic areas (e.g., HIV/AIDS), neglecting other potentially more important domestic priorities." This shows how an over-reliance on donor funding can lead a country to focus on measuring what donors want, rather than what is most critical for its own development.

9.3.3 Strategies for Building a Country-Led M&E System

Transitioning from a donor-dependent to a country-led M&E system requires deliberate and strategic action on the part of the government. This journey is not instantaneous but a gradual process of building ownership, capacity, and self-reliance.

- **Cultivate a Strong Domestic Demand for Evidence:** The most critical first step is for a government to generate its own demand for M&E information. This requires a strong political champion—a President, Minister of Finance, or a senior official—who consistently demands performance data and uses M&E findings for decision-making. When policymakers see the value of M&E as a tool for improving governance, they will be more likely to invest in and own the system.
- **Institutionalize M&E through Policy and Legislation:** Governments must establish a formal national M&E policy and framework that provides a clear mandate for M&E activities and an institutional home for the M&E function. This policy should be binding on all government departments

and should require all development partners to align their M&E activities with the national system. A key example is Uganda's use of the Poverty Reduction Support Credit to successfully "marshal donors toward budget support" and induce a greater reliance on the government's own M&E systems.

- **Allocate Dedicated Domestic Resources:** For a system to be sustainable, it must be funded by the government itself. Governments should allocate a dedicated budget line for M&E in the national budget to fund M&E staff, data collection, and evaluations. This is a tangible demonstration of ownership and a crucial step in reducing donor dependency.
- **Invest in Long-Term Capacity Building:** Building a country-led system requires a long-term investment in the skills of civil servants. This goes beyond short-term training and includes providing opportunities for on-the-job learning, mentoring, and professional development. When government staff have the skills to manage and use their own M&E systems, they are better equipped to lead the process and reduce the need for external technical assistance.

9.3.4 Case Studies of Country-Led M&E Systems in Africa

The experiences of several African countries illustrate the possibilities of a country-led M&E system.

- **South Africa:** The institutionalization of M&E in South Africa provides an excellent example of a country-led model. The M&E system, located in the Presidency, was a product of strong political will and was designed to serve national priorities from the outset. In this case, international partners played a supportive role, providing peer support and knowledge sharing, but the agenda and ownership were firmly in the hands of the government.

- **Uganda:** Uganda's experience is a powerful case study of a country that successfully navigated the transition from a donor-supported to a country-led system. By strategically using its national M&E framework as a basis for donor coordination and by demonstrating a strong political commitment, Uganda was able to reduce the fragmentation of M&E and build a system that is now a core part of its governance.

The journey to M&E institutionalization culminates in the creation of a system that is owned, led, and used by the country itself. The role of external partners, while valuable, must evolve from a dependency-driven relationship to one of genuine partnership and support. By building a strong domestic demand for evidence and investing in sustainable systems, governments can transform M&E from a foreign requirement into a powerful, homegrown tool for national development.

9.4 The Role of Non-State Actors in Demanding and Using M&E

A truly robust and institutionalized M&E system is not a government monopoly. It is an ecosystem that involves a variety of actors who collectively contribute to a culture of evidence-based decision-making. Among the most crucial of these actors are non-state entities, including civil society organizations (CSOs), think tanks, and academic institutions. These groups play a vital role in creating the "demand side" for M&E, ensuring that government performance is not only measured but also scrutinized, verified, and used to promote public accountability. This discussion will delve into the multifaceted role of non-state actors in M&E, exploring how they demand and use

M&E, the dynamics of their relationship with government, and the pathways for building a collaborative and effective M&E ecosystem.

9.4.1 The Crucial Role of Non-State Actors

Non-state actors are essential partners in a national M&E system, performing functions that government institutions often cannot or do not. Their value proposition is rooted in their independence, their proximity to citizens, and their ability to act as a check on government power.

- **As Demands for Accountability:** In many contexts, government M&E systems, particularly in their early stages, are weak or are primarily focused on satisfying donor requirements. In this environment, non-state actors fill a critical vacuum by creating a domestic demand for accountability. By demanding access to M&E data and using it to scrutinize government performance, they incentivize governments to produce credible data and take M&E seriously as a tool for public accountability.
- **As Independent Verifiers:** Non-state actors provide a crucial independent check on the veracity of government-produced M&E data. They can conduct their own data analysis, produce alternative reports, or implement community-based monitoring initiatives to verify whether official M&E findings align with the reality on the ground. This independent verification is essential for building public trust in the M&E system itself.
- **As Facilitators of Public Dialogue:** Government M&E reports are often technical, complex, and inaccessible to the general public. CSOs and think tanks play a vital role in translating these findings into understandable language, thereby fostering a public dialogue about government

performance, policy effectiveness, and resource allocation. This bridges the gap between government-produced data and public understanding, which is a key component of vertical accountability.

9.4.2 How Non-State Actors Demand and Use M&E

Non-state actors employ a variety of strategic approaches to leverage M&E for accountability and policy influence.

- **Community-Based Monitoring (CBM):** CBM is a powerful practice where citizens and communities themselves collect and analyse data to monitor the quality and impact of government services. The Black Sash organization in South Africa provides a perfect example of this. The organization trains "community partners" to conduct on-the-spot surveys at government service sites (e.g., clinics and social security offices) to gather feedback from citizens. The data is then used in dialogues with local government officials to co-create solutions and improve service delivery. This model is a powerful demonstration of how M&E can be used as a tool for citizen empowerment and direct accountability.
- **Policy Analysis and Advocacy:** Think tanks and research institutes are a crucial part of the M&E ecosystem. They use government M&E data and their own independent research to produce policy briefs, reports, and analyses that inform or influence policy. By bringing evidence to the policy-making table, they can challenge assumptions, propose alternative solutions, and push for more effective and equitable policies. They act as a bridge between the world of data and the world of policy, ensuring that M&E findings are not left to gather dust but are used to shape government action.

- **Data Analysis and Public Report Cards:** By accessing government M&E data (e.g., budget allocations, performance reports), CSOs can analyse it and produce their own "report cards" on government performance. These reports, often disseminated through the media and social networks, provide a powerful tool for holding government officials accountable for their performance against their own stated goals. This public-facing approach provides a critical check on government power and helps to mobilize public opinion around key development issues.

9.4.3 Strengthening the M&E Ecosystem: The Government-Civil Society Link

For M&E to reach its full potential, the relationship between government and non-state actors must evolve from one of mutual suspicion to one of collaborative partnership. This requires deliberate effort from both sides.

- **Governments Must Provide Access to Information:** The foundation of this collaboration is transparency. For non-state actors to use M&E effectively, governments must make their M&E data, reports, and findings publicly available in an accessible format. The adoption of Freedom of Information laws and the use of open data portals are essential steps in empowering civil society to play its role.
- **Involve CSOs in the M&E Cycle:** Governments can strengthen their M&E systems by actively involving non-state actors in all stages of the M&E cycle. This can include:
    - **In the design phase:** Involving CSOs in the development of national M&E frameworks and the selection of indicators.
    - **In the data collection phase:** Partnering with CSOs that have a strong presence in communities to gather data and provide

insights from the front lines of service delivery.

- **In the dissemination and use phase:** Including CSOs on evaluation steering committees and using their forums to disseminate M&E findings to the public.

- **Protect the Space for Independent Oversight:** Governments must create an enabling environment for civil society to operate freely and without fear of retribution. This is essential for CSOs to act as a credible and independent check on government performance. A robust M&E ecosystem thrives on open dialogue and constructive criticism, and this can only happen where civic freedoms are protected.

9.4.4 Case Studies from the African Context

Several African countries provide compelling examples of the potential of this government-civil society collaboration.

- **South Africa:** The **Black Sash CBM model** is a shining example of a constructive partnership between civil society and government. While acting as an independent monitor, the organization's approach is collaborative, with the data and findings being used in a dialogue with government officials to co-create solutions. This demonstrates that non-state actors can be both a partner in service delivery and a monitor of government performance.
- **Ghana and Uganda:** These countries have seen deliberate efforts to create collaborative platforms between the government and CSOs in specific sectors. For example, in Ghana, efforts have been made to foster a collaborative M&E platform in the water and sanitation sector. Similarly, in Uganda, the health sector has seen a growing collaboration between the government and CSOs on the evaluation agenda. These examples, though still facing challenges, represent a

crucial step toward building an inclusive M&E ecosystem.

In summary, the role of non-state actors is an indispensable part of a vibrant and effective national M&E system. While governments are responsible for designing and implementing M&E, civil society and think tanks play a crucial role in creating the demand for evidence, holding governments accountable, and ensuring that M&E findings lead to tangible improvements in people's lives. The transition to a collaborative M&E ecosystem is not just a technical change; it is a fundamental shift toward a more transparent, accountable, and democratic model of governance.

Chapter 10

Common Challenges and Best Practices in Africa

Introduction

Monitoring and Evaluation (M&E) across Africa has gained significant traction as governments, development partners, and civil society increasingly recognize its value in driving results and accountability. However, the practice still faces a range of persistent challenges—from limited institutional capacity and weak data systems to political interference and inadequate funding. These obstacles often hinder the effective implementation and utilization of M&E systems. Despite these setbacks, many African countries have also demonstrated innovative solutions and best practices, including the establishment of national evaluation policies, capacity-building initiatives, the use of digital tools, and greater stakeholder engagement. This chapter explores the common hurdles encountered in the African M&E landscape and highlights practical strategies and emerging models that are shaping more responsive and resilient evaluation systems across the continent.

10.1 Synthesis of Challenges: An Interconnected Web

The journey towards institutionalizing a robust Monitoring and Evaluation (M&E) system in African public institutions is often a challenging one. While previous chapters have highlighted the key components of an ideal M&E framework, the reality on the ground is that these efforts are frequently hindered by a complex and interconnected web of obstacles. These challenges are not isolated but rather feed into one another, creating a formidable barrier to progress. This chapter provides a synthesis of four of the most common challenges faced by M&E systems in Africa: financial constraints, political instability and lack of political will, weak data systems, and human capacity gaps. By understanding the nature and interconnectedness of these challenges, governments and their partners can develop more effective and sustainable strategies for M&E institutionalization.

10.1.1 Financial Constraints: The Perennial Challenge

A lack of dedicated and sustained funding is perhaps the most fundamental challenge to institutionalizing M&E. In many governments, M&E is often viewed as a low-priority, "optional extra" rather than a core function of management and governance. This perception has several damaging consequences:

- **Inadequate Budget Allocation:** M&E is rarely allocated a sufficient budget in the national planning and budgeting process. This means that funds for essential M&E activities, such as data collection, large-scale surveys, and independent evaluations, are either non-existent or inadequate. The Journal of Local Government Research and Innovation highlights this problem in South African municipalities, where M&E implementation is often hampered by a lack of resources.
- **Over-reliance on Donor Funding:** The scarcity of domestic funding often leads to an over-reliance on international development partners. While this support can be a valuable catalyst, it also creates a state of donor dependency. As discussed in Chapter 9, this can lead to a fragmented, project-based M&E system that collapses once donor funding is exhausted. The intrac.org platform points out that this over-reliance can prevent governments from developing their own M&E systems, as they are often too busy meeting the specific, and sometimes unhelpful, requirements of various donors.
- **Poorly Resourced M&E Units:** The lack of funding extends to M&E units and departments themselves. These units are often understaffed, lack the necessary software and equipment, and cannot afford to commission independent, high-quality evaluations. This makes it difficult for them to provide credible and timely data to decision-makers.

10.1.2 Political Instability and the Lack of Sustained Political Will

The institutionalization of M&E is fundamentally a political process, and as such, it is highly susceptible to political instability and a lack of sustained political will.

- **The "Politics of M&E":** The notion that M&E is a purely technical and apolitical process is a myth.. As a study in Western Uganda on Community-Based Natural Resources Management projects reveals, M&E is a "political tool serving political objectives." This means that the findings of M&E can be politically sensitive, and there may be a "resistance to transparency" or a "fear of negative findings" among government officials. In such an environment, M&E data may be ignored, manipulated, or simply not collected if it risks exposing poor performance.
- **Lack of Sustained Political Will:** For an M&E system to succeed, it requires a sustained commitment from the highest levels of political leadership. However, political instability—whether due to frequent government reshuffles, electoral cycles, or coups—can easily derail progress. A new administration may not share the M&E vision of its predecessor and may choose to ignore or even dismantle an existing system. The lack of this "magic ingredient" of political will, as the Encyclopaedia of World Problems puts it, can lead to stagnation and inaction.
- **Misalignment with Political Cycles:** The long-term nature of M&E (e.g., an evaluation may take a year to complete) often clashes with short-term political cycles, such as the two-to-four-year electoral cycle. This can lead to a situation where policymakers are more interested in achieving quick, visible

wins for re-election rather than investing in the slow and deliberate process of building a robust M&E system.

10.1.3 Weak Data Systems: The Bedrock of the Problem

An M&E system is only as good as the data it uses. In many African countries, weak national data systems are a fundamental and pervasive challenge that undermines M&E efforts from the very beginning.

- **Rudimentary Data Collection and Quality Issues:** Many government departments and ministries still rely on outdated and inefficient data collection methods, such as rudimentary paper-based monitoring systems and manual data processing. As a case study on project monitoring in Botswana reveals, this leads to significant delays and errors in performance reporting. The lack of standardized data collection protocols also results in poor data quality, making it difficult to generate reliable evidence for decision-making.
- **Lack of Data Interoperability:** Even when data is collected, it often exists in fragmented "silos." Different ministries, departments, and even different donor-funded projects within the same ministry use their own data systems that are not interoperable. The digital health ecosystem in Botswana provides a clear example of this, where "diverse eRecord systems from different vendors and donors exist... and are not interoperable," leading to fragmented care delivery and duplicated effort. This makes it almost impossible to get a holistic picture of government performance across different sectors.
- **Insufficient Analytical Capacity:** Beyond the collection of data, there is often a significant gap in the capacity to analyse it. M&E systems may be good at collecting vast amounts of data, but this data often goes unused because there is a lack

of skilled personnel to turn it into actionable evidence. As a study in Tanzania found, M&E data often "does not meet the needs of stakeholders" because it is not properly analysed and presented for use in planning and budgeting.

10.1.4 Human Capacity Gaps: The Foundation of the System

Even with the best policies and funding, an M&E system cannot function without skilled and motivated people. A lack of human capacity is a major and widespread challenge across African public institutions.

- **Shortage of Qualified M&E Professionals:** There is a general shortage of qualified M&E professionals at all levels of government, from central planning units to line ministries. This often leads to a situation where "monitoring responsibilities are assigned to staff without appropriate training or experience," as highlighted in the **Botswana** case study.
- **High Staff Turnover:** The high turnover rate in the public sector poses a significant threat to building and retaining M&E capacity. When trained M&E staff leave for other opportunities (often in the private sector or international organizations), their knowledge and skills are lost, and the institution is forced to start over. A study in South Africa on an education council, for instance, points to high employee turnover as a major challenge in retaining skilled personnel.
- **Inadequate Training and Professional Development:** While there have been significant efforts in M&E capacity development in Africa, the training provided is often ad hoc and unsystematic. The Development Gateway notes that there is insufficient funding to provide regular staff training, and the training that is provided may not be aligned with the real

needs of the institution, further widening the capacity gap.

10.1.5 Synthesis of Challenges: An Interconnected Web

It is critical to understand that these four challenges are not isolated but are deeply interconnected, forming a vicious cycle that undermines M&E institutionalization.

1. A lack of sustained political will leads to a perception that M&E is unimportant, resulting in financial constraints and a failure to allocate a dedicated budget for M&E.
2. These financial constraints, in turn, make it impossible to hire and retain qualified M&E professionals, leading to significant human capacity gaps.
3. The lack of skilled personnel then makes it difficult to build and manage robust data systems, which remain manual, uncoordinated, and of poor quality.
4. Finally, the absence of credible and timely data, combined with a lack of political demand, ensures that M&E findings are not used for decision-making, reinforcing the perception that M&E is an unproductive function and perpetuating the lack of political will to invest in it.

This interconnected web of challenges highlights the need for a holistic and strategic approach. It is insufficient to address just one of these problems in isolation. A successful M&E institutionalization strategy must tackle all four challenges simultaneously, recognizing their mutual influence and working to break the cycle of underinvestment and underperformance.

10.2 Best Practices for M&E in Africa

Despite the significant challenges, a number of best practices have emerged from the experiences of African countries that have successfully built and sustained their M&E systems. This section outlines some of these key lessons.

10.2.1 Strong Political Leadership and Demand

The most crucial factor for M&E success is a strong and sustained political champion at the highest level of government. The experience of South Africa, with the creation of the Department of Performance Monitoring and Evaluation (DPME) in the Presidency, is a prime example. The DPME's establishment, driven by strong political will, sent a clear signal that M&E was a government priority, not an optional extra. This provided the necessary authority to demand performance data from line ministries and to link M&E findings to planning and budgeting.

10.2.2 Embedding M&E in Core Government Processes

Effective M&E is not a standalone activity; it must be embedded into the core processes of government. This includes:

- **Linking M&E to Planning and Budgeting:** M&E findings must be used to inform the next planning and budgeting cycle. For example, a mid-term evaluation of a national development plan should be used to make adjustments to the plan and reallocate resources where necessary.
- **Harmonizing with Existing Systems:** Rather than creating a new M&E system from scratch, the best practice is to leverage and harmonize with existing government systems, such as public financial management systems, sector-specific information systems, and statistical bureaus.
- **Creating a Clear Institutional Home:** An M&E system needs a clear institutional home with the authority to drive the agenda across the public service. This is often a central agency, such as the Office of the President or Prime Minister, or the Ministry of Finance.

10.2.3 Developing a Phased Approach to Capacity Building

Building M&E capacity is a long-term endeavour that requires a systematic and phased approach.

- **Start with Basic Monitoring:** Instead of trying to build a complex, sophisticated M&E system all at once, it is best to start with basic performance monitoring. This involves setting clear indicators for key government programs and collecting routine performance data.
- **Invest in Long-Term Professional Development:** Capacity building should go beyond short-term training workshops. It requires a long-term investment in professional development, including mentoring, on-the-job training, and the establishment of professional M&E associations. This is vital for building a sustainable pool of M&E professionals.
- **Leverage Partnerships:** Collaboration with external partners, such as CLEAR-AA and local universities, can be instrumental in providing technical assistance and training that is tailored to the needs of the country.

10.2.4 Fostering a Culture of M&E Use and Learning

Ultimately, the success of an M&E system depends on whether its findings are used for decision-making and learning.

- **Disseminate and Communicate Findings:** M&E findings must be actively disseminated and communicated to policymakers, civil servants, the public, and non-state actors. The use of accessible formats, such as public reports and dashboards, is crucial.
- **Create Incentive Structures:** Governments should create incentives that reward the use of evidence and data in decision-making. This can include integrating M&E

performance into civil service performance appraisals.

- **Encourage a Learning Culture:** The M&E system should not be seen merely as a tool for accountability but also as a tool for learning and continuous improvement. This requires creating a safe space for policymakers and managers to discuss negative findings and learn from them without fear of retribution.

In summary, while the challenges to M&E institutionalization in Africa are significant, they are not insurmountable. By learning from the experiences of countries that have succeeded, and by adopting a strategic, holistic approach that addresses the interconnected nature of these challenges, governments can build robust and sustainable M&E systems that drive development and promote good governance.

10.2 Synthesis of Success Factors: A Mutually Reinforcing Cycle

Following the discussion on the formidable challenges to M&E institutionalization in Africa, it is essential to synthesize the key factors that have enabled some countries to overcome these obstacles. The experiences of successful governments and institutions across the continent reveal that a combination of high-level political will, strategic partnerships, and a phased approach to implementation are the critical ingredients for success. These factors are not isolated; they form a mutually reinforcing cycle where each element strengthens the others, thereby creating a sustainable foundation for a national M&E system. This chapter explores each of these success factors in detail, grounding the discussion in real-world examples from the African public sector.

10.2.1 High-Level Political Will: The Non-Negotiable Ingredient

The most critical success factor for institutionalizing M&E is a sustained and unwavering commitment from the highest levels of political leadership. Without this, M&E remains a marginalized and underfunded function. Political will is the foundation upon which all other success factors are built, providing the necessary authority and legitimacy to drive the M&E agenda.

- **Providing Authority and Legitimacy:** When an M&E system is championed by the President, Prime Minister, or a senior cabinet official, it gains the authority to demand data and compliance from line ministries and departments. The establishment of the Department of Performance Monitoring and Evaluation (DPME) in the Presidency of South Africa is a prime example. This strategic placement sent a clear signal that M&E was a government priority, not an optional extra, and gave the DPME the power to hold ministries accountable for their performance.
- **Securing Sustainable Funding:** High-level political will is crucial for securing a dedicated and adequate budget for M&E. When M&E is seen as an essential tool for governance, it is more likely to be protected from budget cuts and prioritized during the national planning and budgeting process. This domestic funding is the bedrock of a sustainable M&E system, reducing the dependency on short-term, donor-driven projects.
- **Protecting Independence:** M&E findings can be politically sensitive, especially when they reveal poor performance. A strong political champion is essential for protecting the M&E system from political pressure and ensuring that its findings are credible and used for decision-making. This political backing provides the necessary safeguard for M&E professionals to produce independent and objective analysis.

10.2.2 Strategic Partnerships: Leveraging External Support for Domestic Gain

While a country-led M&E system is the ultimate goal, strategic partnerships with external actors are invaluable in the institutionalization process. The key is to ensure that these partnerships are aligned with national priorities and strengthen domestic capacity, rather than replacing it.

- **Collaboration with International Development Partners:** International partners can provide crucial technical assistance, funding, and platforms for knowledge exchange. The experience of Uganda demonstrates how a strategic partnership approach can be successful. By using its national development plan as a basis for donor coordination, the government was able to align donor support with its own M&E priorities, thereby reducing the fragmentation of M&E efforts and strengthening its own systems.
- **Engaging with Non-State Actors:** Partnerships with civil society organizations (CSOs), think tanks, and academia are essential for creating a vibrant M&E ecosystem. The African Union's Joint Labour Migration Programme (JLMP) provides a great example of a multi-partner framework where the M&E system was designed through a "participatory process" involving the AUC, donor partners, and civil society representatives. This approach ensures that the M&E system is relevant to a wide range of stakeholders and is more likely to be used for public accountability.
- **Peer Learning and Knowledge Exchange:** Strategic partnerships with regional M&E organizations, such as the African Evaluation Association (AfrEA) and Centres for Learning on Evaluation and Results (CLEAR), provide a

platform for peer learning and knowledge exchange. By learning from the successes and failures of other countries, governments can avoid reinventing the wheel and accelerate their M&E institutionalization journey.

10.2.3 A Phased and Incremental Approach to Implementation

The institutionalization of M&E is a complex and long-term process that cannot be achieved overnight. A "big bang" approach, where a comprehensive and elaborate M&E system is launched all at once, often fails due to a lack of capacity, resources, and political buy-in. A phased and incremental approach, which starts small and builds on early successes, is a more effective and sustainable strategy.

- **Start with the Basics: Focusing on Monitoring:** The most successful M&E systems have often started with basic performance monitoring. This involves setting clear indicators for a few key government programs and systematically collecting routine performance data. This approach is less resource-intensive and allows governments to build capacity and demonstrate the value of M&E before moving on to more complex activities like large-scale evaluations.
- **Building on Early Successes:** A phased approach allows a government to build a culture of M&E gradually. By demonstrating early successes, such as a government department using monitoring data to improve a program, the government can build confidence and generate the necessary political will for further investment. The experiences of countries highlighted in the African M&E Case Studies, including **Ghana and Uganda**, show that an incremental approach is a common and effective practice.
- **Learning and Adapting:** A phased approach is, by its very nature, a process of learning and adaptation. It allows a

government to start with a modest M&E framework, identify what works and what doesn't, and adjust the strategy as it goes along. This pragmatic approach is far more likely to succeed than a rigid, top-down implementation plan that fails to account for the unique political and institutional context of a country.

10.2.4 Synthesis of Success Factors: A Mutually Reinforcing Cycle

The three success factors—political will, strategic partnerships, and a phased approach—are deeply interconnected and form a mutually reinforcing cycle.

1. A strong political will provides the foundation and authority for a government to take the lead in M&E institutionalization.
2. This political leadership then enables the government to forge strategic partnerships with donors and other external actors on its own terms, ensuring that these partnerships are aligned with national priorities.
3. These strategic partnerships, in turn, provide the necessary resources and technical expertise to implement a phased and incremental approach to M&E, allowing the government to start with a strong foundation and gradually build capacity.
4. Finally, the early successes and tangible results generated by a phased approach can strengthen the political will for further investment in M&E, thereby perpetuating the entire cycle.

This synthesis highlights that M&E institutionalization is not a matter of simply implementing a technical system. It is a strategic, political, and long-term endeavour that requires a delicate balance of internal leadership, external support, and a pragmatic approach to implementation.

10.3 Cross-Country Comparisons of M&E Systems in Africa, Highlighting Lessons Learned

The journey toward institutionalizing Monitoring and Evaluation (M&E) is a dynamic and complex process that unfolds differently in each country, shaped by unique political, social, and economic contexts. However, a comparative analysis of these diverse experiences reveals universal lessons that can serve as a roadmap for other African nations embarking on or seeking to strengthen their M&E systems. This discussion synthesizes the experiences of several countries—including South Africa, Uganda, Ghana, and Rwanda—to highlight key success factors, persistent challenges, and actionable lessons that can accelerate the institutionalization of M&E across the continent.

10.3.1 Comparative Analysis of M&E Systems

A brief comparison of M&E systems in a few key African countries provides a useful framework for understanding the lessons learned.

- **South Africa: The Centralized, Political Will-Driven Model:** South Africa's M&E system is distinguished by its strong, top-down institutionalization. The establishment of the Department of Performance Monitoring and Evaluation (DPME) in the Presidency signalled that M&E was a central government priority. The DPME was given the mandate to develop a Government-Wide M&E (GWM&E) Framework, which required all government departments to report on performance against national development goals. The success of this model lies in the high-level political will that provided the necessary authority and funding for the system, making it a powerful tool for accountability. However, as some studies have pointed out, while the system is strong on paper, its effectiveness has sometimes been challenged by political interference and a disconnect with on-the-ground service delivery realities.

- **Uganda: The Phased, Country-Led Transition:** Uganda's experience is a powerful case study in transitioning from a donor-dependent to a country-led M&E system. The journey began with the National Integrated Monitoring and Evaluation Strategy (NIMES) in 2006, which was followed by a comprehensive policy in 2013. The success of this approach lies in its phased nature, allowing the country to gradually build its M&E capacity over a decade. By forging strategic partnerships and aligning donor support with its own priorities, Uganda was able to reduce the fragmentation of M&E and secure a stronger sense of ownership. A key lesson from Uganda is that it is possible to leverage external support as a catalyst for a system that is ultimately driven by national needs.
- **Ghana: The Challenge of Fragmentation and Capacity Gaps:** In Ghana, M&E has been institutionalized at various levels, but it has faced significant challenges. Research has pointed to a lack of standardised M&E practices at the local government level, as well as a lack of specialised personnel. This has led to a situation where M&E efforts are fragmented, with different departments and projects using their own systems and data. This lack of harmonization often results in a poor quality of data and makes it difficult to get a coherent picture of national performance. The lesson from Ghana is that while the intent to institutionalize M&E may be present, without a strong central framework and a concerted effort to build capacity, the system is likely to remain fragmented and ineffective.

- **Rwanda: The Results-Based Management Model:** Rwanda has emerged as a leader in results-based management, with a performance-oriented culture that permeates the public sector. The country's Imihigo system—a performance contract between the President and local government officials—is a powerful tool for accountability that is tied directly to results. The success of this model is rooted in strong political leadership and a culture that demands performance and transparency. However, a key lesson is that this approach requires a dedicated focus on building capacity and ensuring that the data used is credible and reliable.

10.3.2 Key Lessons from Cross-Country Experiences

From these and other country experiences, several key lessons emerge that are crucial for M&E institutionalization efforts across Africa.

- **Lesson 1: Political Will is the Catalyst for All Other Success Factors.** The experiences of South Africa and Rwanda demonstrate that the most successful M&E systems are those that are driven by a strong and sustained political champion. This high-level commitment is the foundation that enables a government to:
    - Establish a clear institutional home for M&E (e.g., in the Presidency or Office of the Prime Minister).
    - Secure a dedicated budget for M&E activities.
    - Demand accountability from line ministries.
    - Protect the independence of M&E from political interference. The lesson is clear: a lack of political will is a common cause for the failure of M&E systems.

- **Lesson 2: A Phased and Adaptive Approach Outperforms the "Big Bang".** The experience of Uganda shows that M&E institutionalization is a journey, not a destination. A phased approach, where a country starts with basic monitoring and gradually builds its capacity for more complex evaluations, is more likely to succeed. This allows a government to build on early successes, learn from its mistakes, and build a culture of M&E without overwhelming the public service. This stands in contrast to countries that have attempted to implement a comprehensive M&E system all at once, only to see it fail due to a lack of resources and capacity.
- **Lesson 3: Harmonization and Alignment are Non-Negotiable.** The fragmentation of M&E efforts due to a multitude of donor requirements is a common challenge, as seen in countries like Ghana. The lesson learned is that for M&E to be effective, there must be a single, national framework that guides all M&E activities. This framework should be designed by the government and require all development partners to align their M&E efforts with national priorities and standards. This harmonization is crucial for reducing administrative burden, improving data quality, and ensuring that M&E findings are used for national decision-making.
- **Lesson 4: The Demand for M&E is as Important as the Supply.** The effectiveness and sustainability of M&E systems are significantly enhanced when there is a strong demand for performance data from a variety of stakeholders, including civil society, the private sector, and academia. The experience of countries with a vibrant civil society, such as South Africa, shows that non-state actors can act as independent verifiers and advocates for accountability, thereby creating a healthy

pressure on the government to produce credible data. The lesson is that governments should not view non-state actors as adversaries but as crucial partners in building a culture of transparency and accountability.

10.3.3 Recommendations for Future M&E Institutionalization Efforts

Based on this cross-country comparison, the following recommendations can guide future M&E institutionalization efforts in Africa:

1. **Secure and Nurture a Political Champion:** Prioritize securing a high-level political champion who can provide the necessary authority and support for M&E institutionalization. This champion must be committed to using M&E findings for decision-making, even when they are negative.
2. **Design M&E as a National System, Not a Donor Project:** Ensure that M&E systems are designed by national experts, tailored to the specific local context, and driven by national priorities. This is crucial for building a sense of ownership and reducing donor dependency.
3. **Invest in Domestic Resources for Sustainability:** Gradually increase the allocation of domestic resources to M&E to ensure that the system can be sustained independently of external funding. This includes budgeting for M&E staff, data collection, and training.
4. **Foster a Collaborative Culture of Learning:** Promote an M&E culture that is centred on learning and continuous improvement, not just on blame and punishment. This requires creating a safe space for dialogue on M&E findings and using the results to improve policy and program

implementation.

The experiences of African countries in institutionalizing M&E are rich with lessons. By learning from the successes of countries like South Africa and Uganda, and by understanding the challenges faced by others, African governments can develop effective and sustainable M&E systems that drive national development and promote good governance.

Chapter 11

Future Directions for M&E Institutionalization in Africa

Introduction

As African nations continue to advance their development agendas, the institutionalization of Monitoring and Evaluation (M&E) emerges as a strategic priority for sustaining impact, enhancing transparency, and strengthening governance. While progress has been made in building national M&E systems, the future demands deeper integration into public sector management, stronger legal and policy frameworks, and the cultivation of a culture of evidence use across all levels of government. Looking ahead, there is also a growing need to harness technology, invest in local capacity, and promote collaborative partnerships among governments, academia, civil society, and international actors. This chapter explores the pathways and opportunities for embedding M&E more firmly into Africa's institutional architecture, offering a vision for systems that are adaptive, inclusive, and capable of supporting transformative change.

11.1 The Link Between M&E and the Sustainable Development Goals (SDGs) and Africa's Agenda 2063

The institutionalization of Monitoring and Evaluation (M&E) in African public institutions has evolved from a donor-driven requirement into a core function of national governance. The future of this institutionalization, however, is not a continuation of the past. It is inextricably linked to two of the most ambitious and transformative development frameworks of our time: the United Nations' Sustainable Development Goals (SDGs) and the African Union's Agenda 2063. These frameworks, which set out a shared vision for a prosperous, peaceful, and sustainable future, have created a powerful new imperative for M&E. They have shifted the focus from measuring project-level results to tracking progress on a vast, interconnected, and globally relevant set of goals and targets. This chapter provides a comprehensive discussion on the crucial link between M&E and these two agendas, exploring why robust M&E is central to their achievement, and identifying the key challenges and opportunities for African nations in this new era.

11.1.1 The New M&E Imperative: The SDGs and Africa's Agenda 2063

The SDGs and Agenda 2063 represent a new paradigm for development, one that is underpinned by a commitment to data, results, and accountability.

- **The Sustainable Development Goals (SDGs):** The SDGs, adopted in 2015, are a "shared blueprint for peace and prosperity for people and the planet, now and into the future." They consist of 17 goals, 169 targets, and a vast set of indicators covering a wide range of development issues, from poverty and hunger to climate action and gender equality. A central tenet of the SDGs is the principle of "leaving no one behind," which places an unprecedented demand for high-quality, disaggregated data to track the progress of the most vulnerable populations.
- **Africa's Agenda 2063:** This is the continent's own "strategic framework for the socio-economic transformation of the continent over the next 50 years." It is a blueprint for a "prosperous and united Africa" based on seven aspirations. Importantly, Agenda 2063 is explicitly results-oriented, with a Continental Results Framework designed to monitor and evaluate implementation at the national, regional, and continental levels. It is a powerful affirmation of Africa's ownership of its own development narrative.

11.1.2 The Crucial Link: Why M&E is Central to These Agendas

M&E is not an optional or ancillary tool for achieving the SDGs and Agenda 2063; it is the central nervous system that connects goals to action and action to results. Without a robust M&E system, these ambitious agendas are little more than aspirations.

- **Evidence-Based Planning and Prioritization:** The sheer number of goals and targets in the SDGs and Agenda 2063 presents a significant challenge for national governments. M&E provides the evidence needed to translate these high-level frameworks into national development plans and to prioritize which goals and targets are most relevant for a country's unique context. It helps to ensure that policies and programs are designed and funded based on what is most likely to have an impact.
- **Tracking Progress and Holding Nations Accountable:** A key purpose of both frameworks is to provide a mechanism for tracking progress and holding nations accountable for their commitments. M&E systems are the vehicles for generating the data and evidence needed to report on progress. Without M&E, there would be no way to know whether a country is on track to achieve its goals, and no way to hold governments, international partners, and other stakeholders accountable for their contributions.
- **Promoting Transparency and Citizen Engagement:** M&E data is a powerful tool for promoting transparency and empowering citizens. By making M&E data on the SDGs and Agenda 2063 publicly available, governments can enable civil society, media, and citizens to engage in the development process, demand accountability, and participate in the monitoring of service delivery. This vertical accountability is essential for building public trust and ensuring that the development process is inclusive.

11.1.3 Challenges for M&E in the New Era

While the SDGs and Agenda 2063 provide a strong impetus for M&E, they also present a new set of formidable challenges for African countries, many of which are already grappling with institutionalization issues.

- **Data Gaps and Disaggregation:** The most significant challenge is the sheer volume of data required to track the hundreds of indicators. The UN has noted that a lack of data for two-thirds of the global SDG indicators remains a major problem. Furthermore, the leave no one behind principle places a strong demand for data to be disaggregated by gender, age, geography, and other factors. African countries, many with weak national statistical systems, face an immense challenge in collecting, analysing, and reporting on this vast and detailed data.
- **Harmonization and Coordination:** A major challenge is the lack of a "harmonized and coordinated M&E system" at the national, regional, and continental levels. African countries must develop a system that can simultaneously track progress on their national development plans, the SDGs, and Africa's Agenda 2063. This requires a high degree of coordination and alignment to ensure that national reporting feeds seamlessly into regional and continental reporting frameworks.
- **Funding and Capacity Gaps:** The ambition of the SDGs and Agenda 2063 is not matched by a proportional increase in funding for the M&E systems required to track them. Countries continue to face inadequate funding, a shortage of skilled M&E professionals, and a lack of investment in the necessary technology and infrastructure. Without a concerted investment in these areas, the ability of African countries to

report credibly on their progress will be severely limited.

11.1.4 Opportunities for M&E in the New Era

Despite the challenges, the SDGs and Agenda 2063 also present an immense opportunity to accelerate M&E institutionalization in Africa and to transform the way development is managed.

- **A New Driver for M&E Institutionalization:** The global and continental commitment to these frameworks provides a powerful new reason for governments to invest in M&E. These agendas can serve as a political catalyst, providing the justification and political will needed to allocate resources, build capacity, and establish a robust national M&E system. The UNDP has argued that the SDGs provide a "strong incentive for governments to strengthen their M&E systems" as a means to achieve their national development goals.
- **Leveraging Technology and Innovation:** The new era of M&E is ripe for technological innovation. The need for real-time, disaggregated data is driving the adoption of new technologies and data sources, such as satellite imagery, big data analytics, and mobile phone data. The Africa Regional Data Cube, for instance, is a platform that uses satellite imagery to provide governments with actionable insights on a range of development issues. These innovations offer a way to overcome some of the traditional data gaps and collection challenges.
- **Strengthening Partnerships and Collaboration:** The SDGs and Agenda 2063 are, by their nature, collaborative, encouraging partnerships between national statistical offices, government ministries, civil society, academia, and international partners. The push to track these goals is creating new platforms and networks for collaboration, which can help

to share knowledge, build capacity, and create a more robust and inclusive M&E ecosystem.

11.1.5 Case Studies and Examples from African Countries

Several African countries are already making significant strides in linking their national M&E systems to these new agendas, providing valuable lessons for others.

- **Egypt and Sierra Leone:** The AUDA-NEPAD has highlighted Egypt and Sierra Leone as countries that are already linking their national development plans and M&E systems to the SDGs and Agenda 2063. This is a concrete example of a strategic approach to aligning national priorities with continental and global frameworks.
- **Kenya, Tanzania, and Ethiopia:** On a more programmatic level, the UN has noted that these countries have made "significant progress" on specific goals, such as access to electricity. These successes are a testament to the fact that their national M&E systems are functioning and generating the data needed to report on progress and drive action.
- **The African Union:** The **African Union** has played a leading role by developing a Continental Results Framework for Agenda 2063 and a Common African Position on the Post-2015 Development Agenda that informed the SDGs. This leadership is providing a harmonized platform for African countries to track and report on their progress.

The institutionalization of M&E in Africa has reached a critical juncture. The SDGs and Agenda 2063 have created a powerful new reason for African nations to invest in robust, country-led M&E systems. While the challenges of data gaps, funding, and harmonization are significant, the opportunity to use these frameworks as a catalyst for institutional change, to leverage technology, and to strengthen partnerships is immense. The future of M&E institutionalization in Africa lies in its ability to effectively serve these twin engines of development.

11.2 Emerging Trends in M&E: Adaptive Management, Complexity-Aware Evaluation, and Decentralized M&E

The institutionalization of Monitoring and Evaluation (M&E) in Africa is a dynamic process, one that is continuously evolving to meet the challenges of modern development. The traditional, rigid M&E frameworks, characterized by linear plans, fixed indicators, and a top-down focus on accountability, are proving to be insufficient in an increasingly complex and unpredictable world. In response, a new generation of M&E approaches is emerging, offering a more flexible, context-specific, and learning-oriented way of working. These trends—adaptive management, complexity-aware evaluation, and decentralized M&E—are not merely theoretical constructs but are being actively piloted and institutionalized across the continent, offering a glimpse into the future of M&E in Africa.

11.2.1 Adaptive Management: Embracing Learning and Flexibility

Adaptive management is a cyclical process of learning and adapting, where program plans are not seen as rigid blueprints but as hypotheses to be tested, refined, and revised based on new evidence and changing circumstances. This approach is a direct response to the limitations of traditional, linear M&E in unpredictable and dynamic contexts.

- **The Rationale for Adaptive Management:** In many development contexts in Africa, cause-and-effect relationships are not always clear, and external factors—such as political shifts, climate change, or volatile markets—can quickly render detailed plans obsolete. In such cases, as noted by SIDA, it is a "waste of time to elaborate detailed plans" that will "quickly become irrelevant." Adaptive management provides a structured way to improve our understanding of a problem through an iterative process of setting objectives, implementing a strategy, and evaluating the outcomes to inform future decisions.

- **Core Components in Practice:** A well-known model for adaptive management is the Collaborating, Learning, and Adapting (CLA) framework used by the United States Agency for International Development (USAID). This framework is built on the idea of creating a culture that encourages questioning, learning, and purposeful adaptation. A key component is the use of fast feedback loops, where managers regularly ask staff about what they have learned and what adaptations they would suggest based on new information.
- **Case Study from South Africa:** The Community Adaptation Small Grants Facility project in South Africa, funded by the Adaptation Fund, provides a concrete example of this approach. The project sought to pilot a new mechanism for local-level climate change adaptation in vulnerable communities. The project was initially approved as a four-year pilot but was extended for a fifth year to accommodate unforeseen delays, demonstrating a flexible approach to implementation. It also used a two-tier process for grant applications, which allowed for continuous learning and adaptation to strengthen project proposals before funding was allocated.

11.2.2 Complexity-Aware Evaluation: Navigating Non-Linear Change

Complexity-aware evaluation is an approach that explicitly recognizes that development outcomes are often the result of non-linear interactions, feedback loops, and unintended consequences. It moves beyond the simplistic search for attribution to a more nuanced exploration of contribution in complex systems.

- **The Rationale for Complexity-Awareness:** Traditional evaluation methods, such as Randomized Control Trials (RCTs), are excellent for establishing a clear causal link between a specific intervention and an outcome in simple or complicated situations. However, in complex contexts—where cause and effect are not clearly understood and there is little agreement among stakeholders—these methods are often insufficient. Complexity-aware evaluation is designed for these situations, which are common in socio-ecological systems and programs that seek to influence social change.
- **Methods in Practice:** Instead of trying to prove attribution, complexity-aware evaluation methods, such as Contribution Analysis, seek to produce a credible, evidence-based narrative of how an intervention has contributed to a change. These methods are typically used alongside a Theory of Change that explicitly sets out how change is supposed to happen, and they involve gathering evidence at different levels to test and refine that theory.
- **Case Study from the Olifants River Basin:** The Resilience in the Limpopo (Olifants Basin) Program (RESILIM-O) in South Africa and Mozambique provides a powerful case study of a complexity-aware, developmental, and learning-centred M&E framework. The program recognized the complex nature of the challenges it was addressing and, instead of

relying on a rigid blueprint, it used a hybrid framework that combined the need for accountability with a strong desire for organizational learning. This involved using new elements such as collaborative case studies and reflection days to facilitate learning among program implementers.

11.2.3 Decentralized M&E: Empowering Local Ownership and Accountability

Decentralized M&E is a trend that involves delegating M&E functions and responsibilities from a central government to sub-national, local, or community levels. It is a fundamental shift that is deeply linked to the broader trend of decentralization in African governance.

- **Benefits of Decentralization:** The core benefit of decentralized M&E is that it increases local ownership of the development process. When local governments and communities are responsible for M&E, the data collected is more likely to be relevant to local needs and to be used for local decision-making. This also enhances vertical accountability, as local officials are held to account by the communities they serve, not just by the central government.
- **Challenges of Decentralization:** The implementation of decentralized M&E is not without its challenges. Chief among these is the significant capacity gap at the local level. Many local governments lack the skilled personnel, funding, and infrastructure to carry out M&E effectively. As a study in South Africa on municipal M&E points out, while there is a strong commitment to M&E, the practice often concentrates on measuring service delivery outputs while neglecting to assess the broader impact due to these capacity constraints.
- **Case Studies from Ethiopia and Ghana:** Ethiopia provides

a clear example of a country that has institutionalized a decentralized M&E system. The country adopted a District Level Decentralization Program (DLDP) in 2002, which devolved power and fiscal responsibilities to the Woreda level (district). The goal was to improve public services and enhance the participation of local communities in political and development affairs. Similarly, in Ghana, the Policy Planning, Monitoring, and Evaluation Divisions at the district level are responsible for M&E activities, demonstrating a commitment to decentralized M&E.

11.2.4 Synthesis of Emerging Trends and Their Relevance to Africa

These three emerging trends—adaptive management, complexity-aware evaluation, and decentralized M&E—are not isolated phenomena but are part of a broader, fundamental shift in M&E practice. They all share a common DNA: a departure from rigid, top-down approaches in favour of systems that are more flexible, more responsive, and more empowering of local actors.

- **Moving Beyond Linear Thinking:** Adaptive management and complexity-aware evaluation both recognize that development is not a linear process with predictable outcomes. They move beyond the simple Plan-Implement-Evaluate model and embrace the reality of complexity and uncertainty.
- **Empowering Local Actors:** Decentralized M&E and adaptive management both place a strong emphasis on the importance of local context and local solutions. Decentralized M&E does this by delegating authority and responsibility to the local level, while adaptive management does so by encouraging a probe-sense-respond approach that is rooted in

local realities.

- **A New Culture of M&E:** Ultimately, these trends point toward a new culture of M&E—one that is more focused on learning and continuous improvement, and less on blame and rigid accountability. They are about creating M&E systems that are not just for reporting to external actors but are also for internal learning and decision-making at all levels of government.

The future of M&E institutionalization in Africa is inextricably linked to the ability of governments to embrace these emerging trends. By moving beyond a rigid, top-down approach and adopting M&E systems that are more flexible, complexity-aware, and decentralized, African nations can build more resilient, responsive, and ultimately more effective development systems that are better equipped to navigate the challenges of the 21st century.

11.3 The Future of M&E as a Tool for Innovation and Policy Entrepreneurship in Africa

The institutionalization of Monitoring and Evaluation (M&E) in Africa has traditionally been framed as a mechanism for accountability—a tool for checking whether programs are on track and for holding managers responsible for performance. While this function remains crucial, the future of M&E in Africa lies in a profound paradigm shift: from a backward-looking tool for control to a forward-looking catalyst for innovation and a powerful resource for policy entrepreneurship. This new role recognizes that in the complex and dynamic landscape of African development, M&E must do more than just report on the past; it must actively inform the future by providing a structured process for learning, adapting, and generating new, evidence-based solutions. This chapter provides a comprehensive discussion on this emerging role of M&E, exploring how it can be a driver of innovation, a tool for policy advocacy, and what is needed to unlock this potential.

11.3.1 M&E as a Catalyst for Innovation

Innovation, in the public sector, is not merely about technological change but about a fundamental shift in how policies, programs, and services are designed and delivered. M&E can serve as a vital engine for this process by creating a space for experimentation, generating evidence on what works, and facilitating institutional learning.

- **Creating a Space for Experimentation:** The rigid, top-down nature of traditional M&E often stifles innovation by discouraging experimentation and risk-taking. A more agile and learning-oriented approach, such as Developmental Evaluation, provides a safe space for policymakers to pilot new ideas on a small scale. By providing real-time feedback, M&E allows innovators to test their hypotheses, learn quickly from failures, and adapt their interventions based on new evidence. The Community Adaptation Small Grants Facility project in South Africa provides a compelling example. By using an adaptive management approach, the project was able to test a new mechanism for local climate change adaptation, gathering crucial evidence on what worked before considering a wider rollout.
- **Generating Credible Evidence for Scaling Up:** For an innovative policy or program to be scaled up, it must have a credible evidence base to prove its worth. M&E provides this crucial evidence. By systematically tracking a pilot program's outcomes, M&E can generate the data needed to demonstrate its tangible impact. This evidence is essential for moving a new idea from a good concept to a proven, evidence-based solution that is worthy of a major investment.

- **Facilitating Institutional Learning:** M&E is a powerful tool for knowledge management and institutional learning. By systematically documenting what was learned from an intervention—both its successes and its failures—M&E helps to build a corporate memory that is essential for innovation. This information can be used to inform the design of new programs and to avoid repeating past mistakes. The RESILIM-O program in the Olifants River Basin is a great example of a program that used M&E to create a hybrid framework that balanced accountability with a strong emphasis on organizational learning, a crucial element for addressing complex, systemic problems.

11.3.2 M&E as a Tool for Policy Entrepreneurship

Policy entrepreneurship is the art of using data, evidence, and political savvy to advocate for new policy ideas and mobilize support for them. In this context, M&E is not just a reporting function; it is a strategic asset for those who seek to drive policy change.

- **Building a Compelling Case with Data:** Policy entrepreneurs need to build a compelling, evidence-based case for their ideas. M&E provides them with the data they need to do this. For example, a civil society organization can use M&E findings to highlight a persistent policy gap, demonstrate the failures of the status quo, and propose an evidence-based alternative. The Twende Mbele initiative's work in Ghana and Uganda provides a great illustration of this, where civil society and government are co-creating a sector-wide evaluation agenda to push for policy change.
- **Demonstrating Impact to Gain Political Support:** For a new policy to gain traction, it must have the support of key stakeholders, including politicians, senior civil servants, and

the public. M&E is a powerful tool for demonstrating the tangible impact of a new policy, thereby building the political will needed to support it. By using M&E to highlight the benefits of a pilot program, policy entrepreneurs can turn a policy idea into a political priority.

- **Empowering a New Generation of Civil Servants:** The future of M&E in Africa lies in its ability to empower a new generation of civil servants to become policy entrepreneurs. By equipping civil servants with M&E skills, we are giving them the tools to not only identify problems but also to develop and advocate for evidence-based solutions. This shifts the role of the civil servant from a passive implementer of policy to an active and engaged driver of policy innovation.

11.3.3 The Enabling Environment for M&E-Driven Innovation

For M&E to fulfil its potential as a tool for innovation and policy entrepreneurship, a supportive enabling environment is required. This is not just a technical requirement but a cultural and political one.

- **A Culture of Learning, Not Blame:** A culture of fear of failure is a significant barrier to innovation. For M&E to be a catalyst for innovation, governments must move beyond a blame-focused culture and embrace a culture of learning, where a project's failure is seen as a learning opportunity rather than a reason for punishment. This requires political leaders and senior managers to create a safe space for experimentation.
- **Political Space for Experimentation:** Political leaders must be willing to give civil servants the space to experiment with new ideas and to accept that not all of them will succeed. This requires a strong, committed political leadership that sees M&E as a tool for building a more effective and responsive

government, as opposed to a tool for political control. The African Evaluation Journal's discussion on the "politics of M&E" in **Uganda** highlights the importance of navigating these political dynamics to unlock the full potential of M&E.

- **Strategic Partnerships with Research Institutions and Civil Society:** The capacity for M&E-driven innovation often extends beyond the government. A key part of the enabling environment is the existence of strategic partnerships between governments and external actors, such as universities, think tanks, and civil society organizations. These partnerships can provide a government with access to new ideas, technical expertise, and an independent perspective that is essential for a robust and innovative M&E ecosystem.

In conclusion, the future of M&E institutionalization in Africa is not about building more rigid systems or collecting more data for the sake of reporting. It is about a fundamental shift in mindset and practice, where M&E is seen as an indispensable tool for innovation and a powerful resource for a new generation of policy entrepreneurs. By embracing agile and learning-oriented approaches, and by fostering an enabling environment that values learning over blame, African nations can build M&E systems that are not just for accountability, but are at the very heart of creating a more effective, responsive, and innovative public service that is capable of delivering on the aspirations of its people.

Chapter 12

Case Studies of Successful Institutionalization

Introduction

To better understand what effective Monitoring and Evaluation (M&E) institutionalization looks like in practice, it is essential to examine real-world examples where countries have made significant strides in embedding M&E into their governance systems. Case studies offer valuable insights into the processes, strategies, and enabling conditions that contribute to sustainable and impactful M&E frameworks. In this chapter, we explore selected African countries that have demonstrated notable progress in institutionalizing M&E, highlighting key drivers of success, innovations adopted, challenges overcome, and lessons learned. These stories not only provide practical guidance for other nations on similar paths but also reinforce the importance of political will, local ownership, and continuous learning in building strong, resilient M&E systems.

12.1 Case Study 1: Uganda's Journey from Donor-Dependency to National Ownership

The institutionalization of Monitoring and Evaluation (M&E) in African governments is a complex and often protracted process. It requires more than just technical expertise; it demands sustained political will, strategic donor engagement, and a fundamental shift in a country's governance culture. While many African nations have struggled to make this transition, a few have emerged as beacons of success, offering invaluable lessons for their peers. This case study delves into the journey of Uganda, a country that has, over the past two decades, systematically transformed its M&E landscape from a fragmented, donor-driven system into a coherent, country-owned framework designed to meet national needs. By examining the key milestones, strategies, and challenges of Uganda's experience, this chapter provides a practical and detailed illustration of the principles discussed throughout this book.

12.1.1 The Context: A Landscape of M&E Overload

In the early 2000s, Uganda's M&E landscape was emblematic of the challenges faced by many developing nations. The proliferation of development aid, while well-intentioned, created a fragmented and uncoordinated system. Each donor, with its own specific reporting requirements and project timelines, demanded a unique set of indicators and data. A World Bank report from this period described the situation as one of "M&E overload," where monitoring had become a "demanding, but unproductive data collection exercise" that primarily served external accountability rather than informing domestic policy.

At the time, the government's M&E capacity was limited and decentralized, with various ministries and departments conducting their own evaluations with little coordination. The result was a glut of unharmonized data, evaluation findings that were rarely used in policy-making, and a public service that was more focused on meeting donor requirements than on generating evidence for its own use. The need for a new approach was clear: Uganda needed to build a national M&E system that was coherent, integrated, and, most importantly, country-owned.

12.1.2 The Strategic Turn: Laying the Foundations for National Ownership

The transition to a country-led system was not a "big bang" event but a deliberate, phased process that began with strong political leadership and the development of foundational frameworks. The journey can be broken down into three key strategic milestones:

- **Milestone 1: Establishing a National Framework with NIMES (2006):** A pivotal step was the development of the National Integrated Monitoring and Evaluation Strategy (NIMES) in 2006. This strategy was the first foundational document to provide a national vision and framework for M&E, addressing the issue of fragmentation head-on. It established a common set of principles, indicators, and reporting protocols, providing a much-needed blueprint for how M&E would be conducted across government.

- **Milestone 2: The Institutional Home and the National M&E Policy (2011/2012):** To give the system political legitimacy and authority, the government strategically assigned the coordination of all M&E activities to the Office of the Prime Minister (OPM). This move was crucial for addressing the issue of "M&E overload" by creating a central hub for coordination. This was followed by the approval of a formal National Policy on Public Sector M&E in 2011/2012. This policy further codified the vision of a results-oriented system, outlining key principles such as evidence-based decision-making, accountability, and managing for results, and it provided the institutional and legal basis for the national M&E system.
- **Milestone 3: Strategic Engagement with Donors through Harmonization:** A key lesson from Uganda's experience is that national ownership does not mean the exclusion of donors. Instead, it means a strategic and assertive approach to donor engagement. The Ugandan government used frameworks such as the Poverty Reduction Support Credit (PRSC) and Sector Wide Approaches (SWAps) to align donor requirements with national priorities. For example, in the Justice, Law, and Order Sector (JLOS), the use of a SWAp helped to replace a "fragmented vertical-based institutional approach" with a harmonized system where pooled funding and a single results framework were the norm. This approach effectively transitioned donors from being fragmented "funders" to being coordinated "partners" who supported a single, country-owned M&E agenda.

12.1.3 Key Success Factors and Lessons Learned

Uganda's journey offers a number of critical lessons for other nations seeking to institutionalize their M&E systems.

- **The Primacy of Political Will:** The most significant factor in Uganda's success was the sustained political commitment from the highest levels of government. The assignment of M&E coordination to the Office of the Prime Minister and the presidential directive that led to the creation of Baraza (citizen-based M&E forums) sent a clear signal that M&E was a national priority. This top-down commitment provided the necessary authority and resources to drive institutional change.
- **A Phased and Incremental Approach:** Uganda's journey was not a rapid revolution but a gradual evolution. The development of NIMES, followed by the M&E Policy, and the phased implementation of sector-specific M&E systems, allowed the government to build capacity and adapt its approach over time. This learning-by-doing approach was far more effective than a "big bang" reform, which often fails due to a lack of institutional readiness.
- **Strategic Use of Donor Partnerships:** By strategically using frameworks like SWAps and the PRSC, Uganda was able to leverage donor resources to strengthen its national system rather than allowing them to fragment it. This demonstrates that a country can be both a beneficiary of aid and the owner of its own M&E agenda, a crucial balance for sustainable institutionalization.
- **Balancing Top-Down and Bottom-Up Accountability:** The introduction of Baraza is a testament to Uganda's commitment to balancing top-down accountability with

bottom-up citizen engagement. These forums created a direct link between local officials and the communities they serve, providing a mechanism for citizens to hold their leaders accountable for service delivery and the use of public funds.

12.1.4 Challenges That Remained and Future Directions

Despite its significant progress, Uganda's M&E system still faces persistent challenges, providing a cautionary tale that the work of institutionalization is never truly finished.

- **Linking M&E to Planning and Budgeting:** A key ongoing challenge is the effective use of M&E evidence to inform planning and budgeting processes. While there is a strong policy framework in place, a study by Twende Mbele noted that the culture of using M&E evidence to inform budget allocations is still not fully entrenched.
- **Capacity Gaps and Data Quality:** While progress has been made, significant capacity gaps remain, particularly at the local government level. Issues such as the "missing middle" of incomplete data, poor feedback loops, and a lack of skilled personnel continue to undermine the quality and utility of M&E findings.
- **Political and Implementation Gaps:** A nuanced challenge is the "implementation gap," where sound policies and M&E findings are sometimes undermined by political expediency or presidential pledges that are not informed by technical advice. This highlights the ongoing need to navigate the politics of M&E and to foster a culture where evidence is valued over political convenience.

In summary, Uganda's journey from a fragmented, donor-dependent M&E system to a country-owned one is a compelling and instructive case study. It demonstrates that with sustained political will, a phased strategic approach, and a commitment to balancing top-down and bottom-up accountability, it is possible for an African nation to build a robust, sustainable M&E system. The lessons learned from Uganda's experience—from the foundational role of national policies to the strategic engagement with donors—are invaluable for any country on a similar path, serving as a powerful reminder that the true potential of M&E can only be realized when it is fully owned and driven by the nation itself.

12.2 Case Study 2: South Africa—The Role of Parliament and Civil Society in M&E

The institutionalization of Monitoring and Evaluation (M&E) is often seen as a technical, administrative function residing within the executive branch of government. However, for M&E to truly drive accountability and good governance, it requires a vibrant "demand side"—a robust ecosystem of actors, including Parliament and civil society, who use M&E information to hold the executive to account. South Africa provides a compelling case study of a country that has, with varying degrees of success, embedded M&E within its broader democratic accountability framework, creating a dynamic and often contentious nexus between the executive, legislative, and non-state sectors. This chapter examines the distinct and collaborative roles of Parliament and civil society in South Africa's M&E system, highlighting the mechanisms they use, the challenges they face, and the critical lessons learned.

12.2.1 The Context: A Constitutional Mandate for Accountability

South Africa's journey in institutionalizing M&E is deeply rooted in its post-apartheid Constitution, which enshrines principles of transparency, accountability, and citizen participation. This constitutional mandate provides the legal and political space for Parliament and non-state actors to play a central role in government oversight. A key institutional development was the establishment of the **Department of Performance Monitoring and Evaluation (DPME)** in the Presidency in 2010. The DPME was tasked with developing the Government-Wide M&E (GWM&E) Framework, which standardized how government departments would plan, monitor, and report on their performance. This provided the crucial "supply side" of performance information that Parliament and civil society would later use as their primary tool for oversight.

12.2.2 The Role of Parliament: Institutionalizing Oversight

Parliament is the primary institutional mechanism for holding the executive to account, and M&E is at the heart of this function. South Africa's Parliament uses several formal mechanisms to exercise its M&E role:

- **Portfolio Committees as the Engine of Oversight:** The most significant oversight happens at the committee level. Each government department is overseen by a dedicated parliamentary portfolio committee that scrutinizes its work. These committees use M&E reports from the DPME, as well as audit reports from the **Auditor-General of South Africa (AGSA)**, to question ministers and senior officials. They hold public hearings, review Annual Performance Plans (APPs) and Annual Reports, and probe departments on issues ranging from school infrastructure delivery to textbook shortages. The Parliamentary Monitoring Group (PMG), a civil society organization, plays a key role in making these proceedings public, thereby facilitating external scrutiny.
- **The Auditor-General's Link to M&E:** The Auditor-General's office is an independent constitutional body that audits government departments and municipalities. Its reports, which assess the efficiency and effectiveness of government spending, provide a crucial financial and

performance lens on government activities. Parliament, through its Public Accounts Committee and other portfolio committees, relies heavily on the AGSA's findings to hold departments accountable for misspent funds and poor performance. This close link between M&E and audit provides a holistic picture of government performance.

- **Challenges to Parliamentary Oversight:** Despite these robust mechanisms, parliamentary oversight is not without its challenges. The Public Affairs Research Institute (PARI) and other studies have highlighted a significant capacity gap among parliamentarians. Many members lack the technical expertise to effectively analyze complex M&E reports and challenge officials on a technical level. Furthermore, political dynamics, including party discipline, can sometimes undermine the objective and non-partisan nature of oversight.

12.2.3 The Role of Civil Society: The "Demand Side" of Accountability

South Africa's civil society has a long and vibrant history, and its role has evolved from a liberation movement to a powerful force for democratic accountability. It acts as the "demand side" of the M&E equation, using government-generated data to hold the executive to account, influence policy, and inform the public.

- **From Data to Advocacy:** Civil society organizations, think tanks, and research institutions leverage M&E data from government reports to build an evidence-based case for advocacy. For example, a coalition of organizations may use a DPME report on service delivery and an AGSA report on public spending to highlight a persistent problem in a specific sector, such as health or education. This evidence is then used to inform media campaigns, public protests, and engagement with policymakers. The Budget Justice Coalition, for instance, is a network of CSOs that uses government M&E data to scrutinize the national budget and advocate for more equitable spending.

- **Informing the Public and Influencing Policy:** The role of civil society is not limited to holding the government to account; it also plays a crucial role in informing the public and shaping policy. Research institutions and think tanks, such as the Inclusive Society Institute, use M&E data to produce independent policy briefs and analyses, thereby contributing to the public discourse and influencing policymakers' decisions.
- **Challenges to Civil Society's Role:** While civil society's role is critical, it also faces significant challenges. Many organizations, especially at the local level, lack the technical capacity to analyse complex M&E data. Furthermore, a lack of trust and a breakdown of communication between government and civil society can be a significant barrier to effective engagement.

12.2.4 The Nexus: A Dynamic Ecosystem of Accountability

The relationship between Parliament, civil society, and the executive is not always confrontational; it is a dynamic ecosystem where synergies are increasingly common.

- **Partnerships in Oversight:** Civil society organizations often partner with parliamentary portfolio committees, providing them with independent analysis, technical expertise, and on-the-ground information to strengthen their oversight function. The PMG acts as a crucial bridge, making parliamentary information accessible to the public and providing a platform for civil society to engage with the legislative process.

- **The Twende Mbele Program:** A key example of a positive nexus is the Twende Mbele program, a partnership between the DPME, the South African Local Government Association, and various civil society organizations. The program aims to co-create M&E tools and evaluations, thereby building a more collaborative and trusting relationship between the demand and supply sides of M&E.

12.2.5 Lessons Learned and Future Directions

South Africa's experience offers invaluable lessons for other countries seeking to embed M&E within a broader democratic framework.

- **Lesson 1: The Importance of the Supply Side:** For Parliament and civil society to play an effective M&E role, there must be a credible and reliable supply of performance information from the government. The DPME's framework, despite its challenges, has provided this essential information.
- **Lesson 2: Capacity Building is a Universal Need:** All actors in the M&E ecosystem—from civil servants to parliamentarians and civil society activists—require continuous capacity building to effectively engage with and use performance information.
- **Lesson 3: The Demand Side is a Vital Partner, Not an Adversary:** The most effective M&E systems are those that embrace a strong demand side. By collaborating with Parliament and civil society, governments can not only enhance accountability but also gain valuable insights and strengthen the legitimacy of their M&E systems.

In summary, South Africa's case study demonstrates that M&E is not just an administrative tool; it is a fundamental pillar of a democratic system. By creating an institutional framework that enables Parliament and a vibrant civil society to use M&E information to hold the government to account, the country has forged a dynamic ecosystem of accountability that is a model for others to emulate.

12.3 Case Study 3: Rwanda—Integrating M&E with Public Financial Management Reforms

The true institutionalization of Monitoring and Evaluation (M&E) is achieved when its findings are no longer seen as a mere reporting function but as a central driver of resource allocation and public financial management (PFM). This integration ensures that budget decisions are not based on historical spending patterns or political expediency, but on credible evidence of what works and what delivers results for citizens. Rwanda provides a compelling and instructive case study of a nation that has, since the early 2000s, systematically and successfully integrated M&E with its PFM reforms. This chapter examines the unique institutional mechanisms, such as the Imihigo system, and the political and cultural factors that have made Rwanda a leader in results-oriented governance, while also providing a balanced view of the system's persistent challenges and criticisms.

12.3.1 The Context: A Post-Conflict Mandate for Results

Rwanda's journey in M&E and PFM integration is deeply rooted in its post-genocide reconstruction efforts. Following the 1994 genocide, the government was faced with the monumental task of rebuilding a devastated nation with a strong mandate for rapid economic growth, accountability, and efficiency. This imperative drove a "paradigm shift" towards a results-oriented culture, where public officials were expected to deliver tangible results that contributed to the national vision. This commitment was codified in the government's foundational national development plan, Vision 2020, which set out ambitious long-term goals for the country. To effectively manage and track progress towards these goals, it was recognized that M&E could not be a separate function but had to be a central pillar of the country's governance and PFM systems.

12.3.2 The Institutional Framework: Linking Performance to the Budget

Rwanda's success in integrating M&E with PFM is a result of a coherent institutional framework that provides clear and enforceable links between performance reporting and budget allocation.

- **Mechanism 1: The Imihigo System:** The most innovative and well-known mechanism is the Imihigo system, a traditional Rwandan practice of performance contracting that was institutionalized in the public sector in 2006. The Imihigo are performance contracts signed annually by public officials, from the President and cabinet ministers down to the mayors of local districts and their staff. These contracts outline specific, measurable, and time-bound targets that are aligned with the national development plan. The UNDP has noted that the Imihigo system is a "powerful tool for driving public sector performance" and a cornerstone of the country's accountability framework. The public nature of the contracts and the annual national performance review ceremony, attended by the President, make it a high-stakes accountability mechanism.
- **Mechanism 2: Performance-Based Budgeting:** Rwanda's budgeting process is designed to be results-oriented. The government has transitioned from traditional input-based budgeting to a performance-based budgeting approach, where ministries and agencies are required to justify their budget requests based on their performance against previously agreed-upon targets. The **Ministry of Finance and Economic**

Planning (MINECOFIN) plays a central role in this process, reviewing the performance data from various M&E systems to inform its budget allocation decisions.

- **Mechanism 3: The Medium-Term Expenditure Framework (MTEF):** The Imihigo system and performance-based budgeting are integrated into a broader PFM framework through the Mid-Term Expenditure Framework (MTEF). The MTEF provides a rolling three-year plan that links Rwanda's national development priorities to the budget. This framework uses M&E data to make informed budget decisions, ensuring that resources are allocated to the sectors and programs that are demonstrating a tangible impact. The Development Gateway has explicitly stated that Imihigo targets are derived from the strategic plans of institutions and that performance information is used to improve budget allocation in the subsequent year.

12.3.3 Key Success Factors and Lessons Learned

Rwanda's experience offers invaluable lessons for other nations seeking to integrate M&E with their PFM systems.

- **Lesson 1: Sustained High-Level Political Will is Essential:** The single most critical success factor has been the unwavering commitment of the country's top leadership. The President's personal engagement in the annual Imihigo performance review, the strong mandate for results, and the willingness to tie budget decisions to performance have created a powerful enabling environment. This leadership has ensured that M&E is not an optional function but a central and non-negotiable part of governance.
- **Lesson 2: A Culture of Performance and a Simple,**

Coherent System: The simplicity and cultural relevance of the Imihigo system have been key to its success. By drawing on a traditional Rwandan practice, the system was more easily understood and institutionalized. The coherence of the M&E-PFM framework—from national vision to Imihigo contracts to budget allocation—has created a strong and unified culture of performance across the public service.

- **Lesson 3: The Primacy of Domestic Ownership:** Rwanda's M&E-PFM integration was not a donor-driven project but a government-led reform that was designed to meet national needs. This domestic ownership has been crucial for the system's sustainability and its ability to withstand external pressures and political changes.

12.3.4 Challenges and Criticisms

While Rwanda's system is widely admired, it is not without its challenges and criticisms, providing a balanced perspective on its limitations.

- **Focus on Measurable Outputs:** A common criticism, as noted in the African Evaluation Journal, is that the Imihigo system can encourage government employees to focus on activities that are easily measurable at the expense of more complex and qualitative outcomes. This can lead to a focus on inputs and outputs rather than on the broader, more transformative impacts that are harder to measure.
- **Top-Down Nature and Lack of Local Autonomy:** The system is largely top-down and driven by a strong central authority. While this has been a key factor in its success, it has also led to criticisms that it may not always foster genuine local-level autonomy or empower local citizens in the decision-making process. The UN has noted the need for

more decentralized M&E practices to ensure findings are used to inform dialogue and decision-making at the local level.

- **Data Quality and Capacity:** Despite the strong institutional framework, challenges persist with data quality and capacity, particularly at the local level. The need for continuous investment in statistical capacity and M&E skills remains a key priority to ensure the credibility and utility of the performance data that drives the system.

In summary, Rwanda's case study provides a powerful and practical example of how an M&E system can be transformed from a reporting function into a central tool for results-oriented governance. By creating a strong institutional framework that integrates M&E with PFM and by fostering an unwavering culture of performance from the highest levels of government, Rwanda has forged a powerful model for driving resource allocation and achieving national development goals. While challenges remain, the lessons learned from its experience are invaluable for any country seeking to build a more accountable, efficient, and results-driven public service.

12.4 Drawing Lessons from Successful African Experiences

The journey toward institutionalizing Monitoring and Evaluation (M&E) is a dynamic and context-specific process, as demonstrated by the diverse experiences of Uganda, South Africa, and Rwanda. While each country's approach was tailored to its unique political, social, and economic landscape, a comparative analysis of their successes reveals a set of universal and actionable lessons. These lessons transcend technical fixes and touch on the fundamental political, institutional, and cultural dimensions of building a robust M&E system. By synthesizing the best practices from these successful African experiences, this chapter provides a consolidated blueprint for other nations seeking to strengthen their governance and accountability frameworks.

12.4.1 Lesson 1: The Primacy of Sustained High-Level Political Will

The single most critical and non-negotiable factor in the successful institutionalization of M&E is sustained political will at the highest levels of government. In each of the case studies, the M&E system's success was directly proportional to the commitment of a political champion.

- **Rwanda:** The unwavering commitment of the President to the Imihigo system provided the necessary authority and accountability. The President's personal involvement in the annual performance review transformed M&E from a bureaucratic exercise into a high-stakes national priority, ensuring that officials at every level were compelled to deliver results.
- **South Africa:** The establishment of the Department of Performance Monitoring and Evaluation (DPME) in the Presidency signalled that M&E was a central government priority. This strategic placement gave the M&E function the political weight and authority to demand compliance from other government departments, thereby creating a coherent, government-wide framework.
- **Uganda:** The sustained political support for the National Integrated Monitoring and Evaluation Strategy (NIMES) and the strategic placement of M&E coordination in the Office of the Prime Minister gave the system the political legitimacy it needed to overcome early fragmentation and bureaucratic inertia.

The lesson is clear: without a political champion at the top who not only supports M&E but actively uses its findings for decision-making, even the most technically sound system is destined to fail.

12.4.2 Lesson 2: The Power of Phased and Context-Specific Institutionalization

The successful institutionalization of M&E is a long-term journey, not a short-term project. The countries that succeeded avoided a "big bang" or "one-size-fits-all" approach, instead building their systems incrementally and tailoring them to their unique cultural and institutional contexts.

- **Uganda:** Uganda's phased journey, from the development of NIMES in 2006 to the approval of its national M&E policy in 2011/12, allowed the government to build capacity, learn from early pilots, and adapt its approach over time. This incremental strategy fostered ownership and sustainability.
- **Rwanda:** The genius of Rwanda's Imihigo system was its cultural relevance. By adopting a traditional practice of public performance contracts, the government made the M&E system more accessible, understandable, and deeply owned by the populace. This proved far more effective than importing a rigid, foreign-designed framework.
- **South Africa:** The Government-Wide M&E (GWM&E) framework was designed to fit within the country's unique constitutional and political landscape, particularly the need to balance the authority of the executive with the oversight role of Parliament. This customization made the system more legitimate and enforceable.

The lesson is that M&E systems must be homegrown, not imported. They should be built incrementally, allowing for a process of "learning by doing" that fosters a sense of ownership and adaptability.

12.4.3 Lesson 3: M&E as a Tool for Integration, Not Isolation

The true value of M&E is unlocked when it is not an isolated reporting function but is seamlessly integrated into the core functions of government, particularly public financial management and national planning. This integration ensures that M&E findings have a direct impact on resource allocation and strategic decision-making.

- **Rwanda:** Rwanda provides the clearest example of this integration. Its M&E system, through the Imihigo framework, is directly linked to the Public Financial Management (PFM) process. Performance data is used to inform budget allocations, making M&E a central driver of resource management and national development.
- **Uganda:** Uganda's strategy involved using M&E as a central component of its national development plans, such as the Poverty Eradication Action Plan (PEAP). It also strategically used M&E to harmonize donor reporting through Sector-Wide Approaches (SWAps), thereby integrating a multitude of M&E requirements into a single, national framework.
- **South Africa:** The DPME's role was to provide a common language and set of performance information for all government departments. This allowed for M&E findings to be integrated into the planning processes of every ministry and agency, ensuring that performance data was not confined to a single report but was part of the entire governance ecosystem.

The lesson is that M&E's potential is fully realized when it moves beyond a reporting function and becomes an integral part of the machinery of government, from policy formulation to budget execution.

12.4.4 Lesson 4: The Crucial Role of the "Demand Side" of Accountability

A robust M&E system requires more than just a "supply" of data from the executive branch. It requires a strong "demand side" from

Parliament, civil society, and citizens who use M&E information to hold the government accountable.

- **South Africa:** South Africa provides a model for a strong demand side. It's Parliament, through a system of powerful portfolio committees, that uses M&E reports from the DPME and the Auditor-General to rigorously scrutinize the executive. Furthermore, a vibrant civil society, think tanks, and the media use this publicly available M&E data to inform public debate, advocate for policy change, and hold the government to account.
- **Uganda:** Uganda's commitment to citizen engagement is demonstrated through its Baraza system, where local officials meet directly with community members to report on the use of public funds and account for their performance. This mechanism brings M&E and accountability closer to the people.
- **Rwanda:** The public and transparent nature of the Imihigo contracts, which are often displayed in public spaces, fosters a form of social accountability, as citizens can track the progress of their local leaders against publicly declared targets.

The lesson is that M&E is not a tool for the government alone. It is a powerful tool for all of society, and its legitimacy is enhanced when it is used to foster dialogue and accountability between the state and its citizens.

12.5 A Blueprint for the Future

The successful experiences of Uganda, South Africa, and Rwanda provide a clear blueprint for the future of M&E institutionalization in Africa. They show that while the journey is challenging, it is not an insurmountable task. The key to success lies in a consolidated approach built on four core pillars:

1. **Secure Political Leadership:** Identify and nurture a political champion at the highest level of government who is committed to using M&E for decision-making.
2. **Embrace a Homegrown Approach:** Resist the temptation of imported blueprints and instead build a context-specific M&E system incrementally, allowing for continuous learning and adaptation.
3. **Strategically Integrate M&E:** Ensure that M&E is not an isolated function but is seamlessly integrated into core government processes, particularly national planning and the budget.
4. **Foster Broad-Based Engagement:** Build a strong "demand side" for M&E by empowering Parliament, civil society, and citizens to use performance information to hold the government accountable.

By following this blueprint, African nations can build M&E systems that are not just for reporting on the past but are at the heart of building a more efficient, transparent, and results-driven future for the continent.

www.ingramcontent.com/pod-product-compliance
Lightning Source LLC
LaVergne TN
LVHW090319210926
848480LV00001B/293